Optimization of Biological Sulphate Reduction to Treat Inorganic Wastewaters: Process Control and Potential Use of Methane as Electron Donor

THESIS COMMITTEE

THESIS PROMOTOR

Prof. dr. ir P.N.L. Lens
Professor of Biotechnology
UNESCO-IHE, Delft, The Netherlands

THESIS CO-PROMOTOR

Dr. G. Esposito,
Associate Professor of Sanitary and Environmental Engineering
University of Cassino and Southern Lazio, Cassino, Italy

Dr. ir. Yu ZHANG
Laboratory of Microbial Oceanography
Shanghai Jiao Tong University, Shanghai, China

Dr K.J. Keesman
Associate professor, Biomass Refinery and Process Dynamics
Wageningen University, The Netherlands

Dr L. Frunzo
Assistant Professor in Applied Mathematics
University of Naples "Federico II", Naples, Italy

Dr H. Lubberding
Senior Lecturer in Microbiology
UNESCO-IHE, Delft, The Netherlands

OTHER MEMBERS

Prof. V. Belgiorno, PhD
Professor of Civil and Environmental Engineering
University of Salerno, Salerno, Italy

Dr. R. Gerlach, PhD
Associate Professor of Chemical and Biological Engineering
Montana State University, Bozeman, USA

Dr. E. van Hullebusch, PhD
Associate Professor of Biogeochemistry
University of Paris-Est, Institut Francilien des Sciences Appliqueées
Champs sur Marne, France

This research was conducted under the auspices of the Erasmus Mundus Joint Doctorate Environmental Technologies for Contaminated Solids, Soils, and Sediments (ETeCoS^3).

Joint PhD degree in Environmental Technology

Docteur de l'Université Paris-Est
Spécialité : Science et Technique de l'Environnement

Dottore di Ricerca in Tecnologie Ambientali

Degree of Doctor in Environmental Technology

Thèse - Tesi di Dottorato - PhD thesis

Joana Cassidy

Optimization of Biological Sulphate Reduction to Treat Inorganic Wastewaters: Process Control and Potential Use of Methane as Electron Donor

Defended on December 17th, 2014

In front of the PhD committee

Prof. Robin Gerlach	Reviewer
Prof. Vincenzo Belgiorno	Reviewer
Prof. Dr. Ir. Piet Lens	Promotor
Prof. Stéphanie Rossano	Promotor
Prof. Giovanni Esposito	Co-promotor
Prof. Eric van Hullebusch	Examiner

European Commission
ERASMUS
MUNDUS

Erasmus Joint doctorate programme in Environmental Technology for Contaminated Solids, Soils and Sediments (ETeCoS^3)

CRC Press/Balkema is an imprint of the Taylor & Francis Group, an informa business

© 2015, Joana Cassidy

Published by:
CRC Press/Balkema
PO Box 11320, 2301 EH Leiden, The Netherlands
e-mail: Pub.NL@taylorandfrancis.com
www.crcpress.com – www.taylorandfrancis.com

ISBN 978-1-138-02950-7 (Taylor & Francis Group)

TABLE OF
CONTENTS

CHAPTER 1

CHAPTER 2

CHAPTER 3

SULPHIDE RESPONSE ANALYSIS FOR SULPHIDE CONTROL
USING A pS ELECTRODE IN SULPHATE REDUCING BIOREACTORS

CHAPTER 4

BIOPROCESS CONTROL OF SULPHATE REDUCTION IN AN INVERSE FLUIDIZED
BED REACTOR: ROLE OF MICROBIAL ACCUMULATION AND DYNAMIC
MATHEMATICAL MODELLING

CHAPTER 5

EFFECT OF ALTERNATIVE CO-SUBSTRATES ON THE RATE OF
ANAEROBIC OXIDATION OF METHANE AND SULPHATE REDUCTION

CHAPTER 6

GENERAL DISCUSSION AND RECOMMENDATIONS

LIST OF FIGURES

LIST OF TABLES

ACKNOWLEDGEMENTS

I would like to take this moment to say thank you to all the fantastic people who have helped me during these three years of PhD.

First of all I would like to say thank you to the committee members of the Environmental Technologies for Contaminated Solids, Soils and Sediments (ETeCoS^3): Prof. Piet Lens (UNESCO-IHE institute, The Netherlands), Dr. Hab. Giovanni Esposito (University of Cassino and the Southern Lazio, Italy) and Dr. Hab. Eric van Hullebusch (University of Paris-Est, France) for believing in my ability to participate in this incredible program.

I would like to say thank you to my promotor Piet Lens for all the scientific discussions, crucial suggestions, critical views and commitment. All this taught me how to have a more critical scientific thinking and was indispensable for the conclusion of this PhD. I would also like to say thank you to my supervisor Henk Lubberding at UNESCO-IHE for his invaluable support throughout all the difficulties inherent to doing a PhD and for all our long conversations (both scientific and not scientific). Thank you for helping me develop further my scientific thought and for all the insight in the amazing world of microbiology. Thank you to Karel Keesman for his valuable comments and suggestions. I would also like to recognize and appreciate all the lab technicians for all their help and support.

Many thanks to my co-promotor Giovanni Esposito for his support in all different matters regarding the PhD, especially during my stay in Italy, and for having taught me so much about mathematical modelling and its applications. Still regarding my stay in Italy, I would like to acknowledge Dr Luigi Frunzo for his crucial help with Matlab® and mathematical model implementation, and to everyone at the department of Hydraulic, Geotechnical and Environmental Engineering of the University of Naples.

A very special thank you to Yu Zhang and to everyone in her laboratory at Jiao Tong Shanghai University. What an experience! Thank you for making me feel at home in a place so far away with such a different and amazing culture! It was a great experience, both personal and professional. Thank you for allowing me to work in your lab, for all the discussions, the support and knowledge.

Thank you to all the PhD and non-PhD colleagues/friends with whom I shared these 3 years in Holland, Italy and China. It has been a great pleasure to share this multicultural experience with you all. It was a real pleasure to walk this road with you. I am sure our paths will cross in the future.

And last, but definetely not the least I would like to thank my marvelous friends for all the support and for always believing in me. To my amazing boyfriend who is always there for me, for all his help, love and support. To my wonderful family in England and Portugal. To my grandparents. To my mother and sister who are my everything. And to my father and to my grandmother Olga, who I am sure are so proud and watching over us from wherever they are.

SUMMARY

This work investigated two different approaches to optimize biological sulphate reduction: to develop a process control strategy to optimize the input of an electron donor and to study how to increase the feasibility of using a cheap carbon source such as methane.

For the design of a control strategy that uses the organic loading rate (OLR) as control input, feast and famine behaviour conditions were applied to a sulphate reducing bioreactor to excite the dynamics of the process. Such feast/famine regimes were shown to induce the accumulation of carbon, and possibly sulphur, storage compounds in the sulphate reducing biomass. This study showed that delays in the response time and a high control gain can be considered as the most critical factors affecting the application of a sulphide control strategy in bioreactors. The delays are caused by the induction of different metabolic pathways in the anaerobic sludge including the accumulation of storage products.

Polyhydroxybutyrate (PHB) and sulphate were found to accumulate in the biomass present in the inversed fluidized bed bioreactor used in this study, and consequently, they were considered to be the main storage compounds used by SRB. On this basis, a mathematical model was developed which showed a good fit between experimental and simulated data giving further support to key role of accumulation processes.

In order to understand the microbial pathways in the anaerobic oxidation of methane coupled to sulphate reduction (AOM-SR) diverse potential electron donors and acceptors were added to *in vitro* incubations of an AOM-SR enrichment at high pressure with several co-substrates. The AOM-SR was stimulated by the addition of acetate which has not been reported for any other AOM-SR performing communities. In addition, acetate was formed in the control group probably resulting from the reduction of CO_2. These results support the hypothesis that acetate may serve as an intermediate in the AOM-SR process, at least in some groups of anaerobic methanotrophic archaea (ANME) and sulphate reducing bacteria.

RÉSUMÉ

Ce travail a étudié deux approches différentes pour optimiser la réduction biologique des sulfates : la première approche consisté à élaborer une stratégie de contrôle de processus pour optimiser l'ajout d'un donneur d'électrons et la deuxième à vérifier la pertinence de l'utilisation d'une source de carbone bon marché, à savoir, le méthane.

Une stratégie de contrôle de l'apport du donneur d'électron en se basant sur le suivi de la charge organique à été mis en place. Des conditions d'abondance et de famine ont été appliquées à un bioréacteur à bactéries sulfato-réductrices (BSR) pour stimuler les dynamiques du processus. Ces conditions d'abondance/famine ont donnée lieu à l'accumulation de carbone et également de soufre élémentaire (composants de stockage de biomasse réductrice de sulfate). Cette étude a montré que les retards dans le temps de réponse et un gain de commande élevé peuvent être considérés comme les facteurs les plus critiques affectant l'application d'une stratégie de contrôle de sulfure dans des bioréacteurs à BSR. L'allongement du temps de réponse est expliqué par l'induction de différentes voies métaboliques au sein des communautés microbienne des boues anaérobies, notamment par l'accumulation de sous produits de stockage. Le polyhydroxybutyrate (PHB) et les sulfates ont été retrouvés accumulés par la biomasse présente dans le bioréacteur à lit fluidisé inverse utilisé pour cette étude et donc ils ont été considérés comme les produits majoritaires de stockage par les BSR. Sur cette base, un modèle mathématique a été développé, qui montre un bon compromis entre les données expérimentales et simulées, et confirme donc le rôle clé des processus d'accumulation.

Afin de comprendre les voies métaboliques impliquées dans l'oxydation anaérobie du méthane couplé à la réduction des sulfates (AOM-SR), différents donneurs et accepteurs d'électrons ont été ajoutés au cours de test d'incubations *in vitro* visant à enrichir la communauté microbienne impliqué dans l'AOM-SR à haute pression avec plusieurs co-substrats. L'AOM-SR est stimulée par l'addition de l'acétate ce qui n'a pas été rapporté pour d'autres communautés impliqué dans l'AOM-SR. En outre, l'acétate a été généré dans le test de contrôle résultant probablement de la réduction de CO_2. Ces résultats renforcent l'hypothèse que l'acétate peut servir d'intermédiaire dans le processus de l'AOM-SR, au moins pour certains groupes de archées anaérobie méthanotrophe (ANME) et les bactéries sulfato-réductrices.

SOMMARIO

Nel presente studio, due differenti approcci per ottimizzare la riduzione biologica dei solfati sono stati esaminati: sviluppare una strategia di controllo del processo per ottimizzare il dosaggio del donatore di elettroni, e studiare come aumentare la fattibilità dell'utilizzo di una fonte di carbonio a basso costo come il metano.

Per progettare una strategia di controllo che utilizzi il fattore di carico organico come controllo in ingresso, sono state applicate condizioni di abbondanza e privazione a un bioreattore solfato-riducente , in modo da favorire le dinamiche del processo. Tali regimi di abbondanza/privazione hanno mostrato buone potenzialità nel causare l'accumulo dei composti carboniosi, e in alcuni casi di quelli dello zolfo, nella biomassa solfato-riducente. Il presente studio ha mostrato che i ritardi nel tempo di risposta ed un elevato aumento del controllo possono essere considerati come i fattori maggiormente critici nell'influenzare l'applicazione di una strategia di controllo del solfuro nei bioreattori. Tali ritardi sono causati dall'induzione di differenti vie metaboliche nel fango anaerobico, tra cui l'accumulo di prodotti di stoccaggio. In particolare, è stato osservato l'accumulo di poliidrossibuttirato (PHB) e solfati nella biomassa presente nel reattore a letti fluidizzati inversi usato in questo studio, e di conseguenza, questi sono stati considerati come i principali composti utilizzati dai batteri solfato-riduttori. Su queste basi è stato sviluppato un modello matematico che ha mostrato una buona correlazione tra i dati sperimentali e quelli simulati, dando ulteriore supporto al ruolo chiave dei citati processi di accumulo.

Per comprendere le trasformazioni biochimiche alla base dell'ossidazione anaerobica del metano accoppiata alla solfato-riduzione (AOM-SR), diversi potenziali donatori e accettori di elettroni sono stati aggiunti a incubazioni *in vitro* di un AOM-SR a pressione elevata arricchito con diversi co-substrati. L'AOM-SR è stato stimolato dall'aggiunta di acetato, il che non è stato riportato per nessun'altracomunità in grado di svolgere l'AOM-SR. Inoltre, nel gruppo di controllo si è formato acetato, probabilmente a causa della riduzione di CO_2. Questi risultati sostengono le ipotesi che l'acetato possa agire da intermediario nel processo AOM-SR, almeno per quanto riguarda alcuni gruppi di archaea anaerobici metanotrofici (ANME) e batteri solfato-riduttori.

SAMENVATTING

In deze studie worden twee verschillende benaderingen onderzocht om biologische sulfaatreductie te verbeteren: het ontwikkelen van een procescontrole strategie om de toevoer van electrondonoren te optimaliseren en het onderzoeken van de haalbaarheid van goedkope koolstofbronnen zoals methaan. Om de organische belasting (OLR) als invoercontrole te kunnen gebruiken in deze procescontrole strategie werd de sulfaatreducerende bioreactor onderworpen aan "feast and famine" omstandigheden om de dynamiek van het proces te maximaliseren. Dit "feast/famine" regime had als gevolg dat er koolstofverbindingen – en mogelijk ook zwavel – als reservestof werden gevormd in de sulfaatreducerende biomassa. Deze studie laat zien dat vertraging van de reactietijd en een "high control gain" beschouwd kunnen worden als de meest kritische factoren in de toepassing van een sulfide controle strategie in bioreactoren. De vertraging werd veroorzaakt door de inductie van verschillende metabole processen in het anaerobe slib, zoals de ophoping van reserveproducten. Polyhydroxybutyraat (PHB) en sulfaat waren de voornaamste verbindingen die opgehoopt werden in de biomassa van de "inversed fluidized bed" reactoren die in dit onderzoek gebruikt werden en deze werden dan ook beschouwd als de voornaamste reserveproducten van de sulfaatreducerende bacteriën. Op grond van deze resultaten is een mathematisch model ontwikkeld waarin de experimentele en gesimuleerde data uitstekend overeen kwamen, hetgeen de sleutelrol van dit ophopingsproces verder aannemelijk maakt.

Om de microbiologische processen in de anaerobe oxidatie van methaan gekoppeld aan sulfaatreductie (AOM-SR) beter te kunnen begrijpen werden meerdere electronendonoren en –acceptoren gebruikt in *in vitro* incubaties van AOM-SR ophopingen onder hoge druk en met verscheidene cosubstraten. De AOM-SR populatie werd gestimuleerd door de toevoeging van acetaat, wat niet eerder werd waargenomen in AOM-SR populaties. Verder werd acetaat gevormd in de controles, waarschijnlijk als gevolg van de reductie van CO_2. Deze resultaten ondersteunen de hypothese dat acetaat dient als tussenproduct in het AOM-SR proces, tenminste in sommige combinaties van anaerobe methanotrofe archaea (ANME) en sulfaatreducerende bacteriën.

1

Introduction

1.1 PROBLEM STATEMENT

Sulphate and other sulphur compounds (thiosulphate, sulphite, sulphide and dithionite) are common contaminants of fresh water from the release of industrial activities such as the production of edible oil, tannery, food processing, fermentation industry, coal mining and paper/pulp processing (Shin et al., 1996). Although the environmental risk associated with sulphate is comparatively less when compared to other pollutants, sulphate pollution is a concern as it increases the salinity in fresh water. In the absence of oxygen and nitrate, sulphate reduction by anaerobic microorganisms causes an increase in hydrogen sulphide which is toxic and causes an unpleasant smell and corrosion problems (Sawyer et al., 2003). The biological sulphate reduction (SR) process is mediated by a group of microorganisms known as sulphate reducing bacteria (SRB). Biological anaerobic reduction of sulphate has been successfully applied for the treatment of sulphate contaminated wastewater from industries on a larger scale for many years as it offers the possibility of an efficient treatment with low operation costs using various organic and easily utilizable electron donors (Liamleam and Annachhatre, 2007).

Further optimization of the process is required in order to reduce costs. As electron donors are one of the major costs for SR, this research focused on how to optimize its input. This work investigated two different approaches to optimize biological sulphate reduction: the development of a process control strategy to optimize the input of a commonly used electron donor, i.e., lactate, and how to increase the feasibility of using a cheap carbon source such as methane.

Methane as an electron donor for biological sulphate removal in wastewater treatment presents as main advantages its availability and decreased cost as compared to the commonly used electron donors for SR. In addition, the use of methane would close its cycle of utilization, decrease the emission of one of the most important greenhouse gases and reduce the risk of excess carbon source in the treated effluent. The main bottleneck of this process is the extremely low growth rates of the microorganisms responsible for the conversion of methane and of sulphate to sulphide (Meulepas et al., 2010b). Thus, this research tried to get further insight on how to increase the feasibility of using methane for the treatment of sulphate containing wastewater, by comparing potential electron donors and acceptors to *in vitro* incubations of anaerobic oxidation of methane coupled to sulphate enrichment at high pressure with several co-substrates.

In the sulphate reducing process, bioprocess control can be used to regulate the competition between microbial groups, to optimize the input of the electron donor and/or to maximize or minimize the production of sulphide. Controlling the production of sulphide in a sulphate reducing bioreactor is highly relevant to avoid overproduction of sulphide as it increases operational costs and may impose a sulphide removal post-treatment step (Villa-Gomez et al., 2014). This work focused on key aspects for the development of a bioprocess control strategy, i.e., the use of a suitable sensor for sulphide monitoring, the design of an adequate control strategy and control parameters, and process model development.

1.2 BIOTECHNOLOGICAL ASPECTS OF SULPHATE REDUCTION

There are several methods for the removal of sulphate from wastewater including membrane filtration, chemical methods and biological methods. The first two are expensive, and require a post-treatment of the brine. For high-strength sulphate containing wastewaters, biological sulphate removal is a cost-effective alternative (Lens et al., 1998). Biological desulfurization is performed by a group of bacteria: SRB. These bacteria are classified into two subgroups: autotrophic and heterotrophic SRB. Heterotrophic SRB use organic compounds as substrate and autotrophic SRB use CO_2 as carbon source and H_2 as an electron source (Liamleam and Annachhatre, 2007). Biological anaerobic reduction of sulphate has been successfully applied for the treatment of sulphate contaminated wastewater from industries on a larger scale for many years as it offers the possibility of an efficient treatment with low operation costs using various organic and easily utilizable carbon sources (Liamleam and Annachhatre, 2007). The end product is hydrogen sulphide, hence this technique is suitable for treatment of water containing dissolved metals which can be precipitated simultaneously and removed as stable precipitates of sulphide.

Numerous bioreactor systems have been used for sulphate reduction (Figure 1.1). The activity of a bioreactor is given by the activity and concentration of the biomass. Thus, biomass retention is essential to increase its concentration due to the low growth rate of anaerobic microorganisms responsible for the conversion.

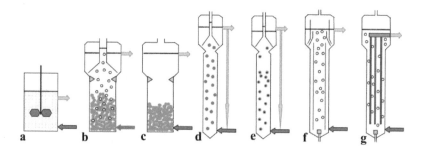

FIGURE 1.1 Bioreactor systems for sulphate reduction. a) continuous stirred tank reactor; b) upflow anaerobic granular sludge bed with b) gas production and c) without gas production; d) expanded granular sludge bed; e) fluidized bed reactor; f) gas-lift bioreactor; g) submerged membrane bioreactor (Bijmans, 2008).

Most wastewaters from industries are deficient in dissolved organic matter therefore, supplementation of electron donors to the sulphate reduction system is needed. Electron donors most commonly used are hydrogen, CO, formate, methanol, ethanol, lactate and acetate. The application of the various electron donors for sulphate removal from various types of wastewater has been extensively reviewed (Liamleam and Annachhatre, 2007) with hydrogen being the most commonly used electron donor. However, lactate is reported as the best suited carbon source as many species of sulphate reducers can use it (Postgate, 1984; Koydon, 2004). Acetate is a key intermediate in the breakdown of organic substances in anaerobic processes and can be used as an electron donor in the sulphate reduction process. However, when incomplete oxidizing sulphate reducers are present, acetate will be not utilized. Acetate production during the biological sulphate reduction is actually a major drawback of sulphate reducing reactors because acetate producing SRB cannot completely oxidize acetate even with excess sulphate levels (Lens et al., 2002). More recently, methane was also found to be a possible electron donor for SR processes (Nauhaus et al., 2002). Several aspects need to be considered when choosing the most appropriate electron donor: price, availability, residual colour or pollution, suitability for a specific waste or process water (volume, composition and salinity) and legislation regarding safety and environment (Bijmans, 2008).

1.3 BIOPROCESS CONTROL OF SULPHIDE PRODUCTION IN BIOREACTORS

For metal removal and recovery processes, the required amount of sulphide to be produced by SRB depends on the composition of the wastewater to be treated, i.e., its metal concentration. Steering the sulphide production towards this required stoichiometric amount in bioreactors is highly relevant to avoid over sulphide production which increases operational costs and may impose a sulphide removal post-treatment step (Villa-Gomez et al., 2014).

A proportional-integral-derivative (PID) controller has been successfully used in anaerobic bioreactors, partly due to the derivative control parameter used to overcome the lag presence caused by the time required for substrate degradation prior to being used or transformed to the desired product (Dunn et al., 2005; Jagadeesh and Sudhaker, 2010). The PID controller has three adjustable parameters (Kc or controller gain, τi or integral time and τd or derivative time) which values can be obtained by using different tuning methodologies and tested experimentally or through model simulations (Pind et al., 2003).

A well-tuned PID controller has parameters which are adapted to the dynamic properties of the process, so that the control system becomes fast and stable. If the process dynamic properties vary without the controller being re-tuned, the control system will have less stability and may become more sluggish. The problems encountered with variable process dynamics can be solved by tuning the controller in the most critical operation (conservative tuning) so that when the process operates in a different operation point, the stability of the control is better. However, if the stability is too good the tracking speed is reduced giving more sluggish control (Cassidy et al., 2015). Another option to solving these problems is to use adaptive tuning, in which the controller parameters are varied along with variations of the process dynamics, so that the performance of the control system is maintained or optimized at any operation point. Typically, adaptive optimal control seeks for a maximum in a performance index function (Heinzle et al., 1993) and requires a model that accounts for changing process conditions. (Further information on this topic is given in Chapter 2 of this thesis).

1.4 ANAEROBIC OXIDATION OF METHANE COUPLED TO SULPHATE REDUCTION

Methane is a potent greenhouse gas which can drastically change the earth climate. In addition, high levels of sulphate in fresh water and marine environments are undesirable as it increases the salinity. Thus, anaerobic oxidation of methane coupled with the sulphate reduction (AOM-SR) process in anoxic environments, plays an important role in controlling the earth's climate and marine ecosystems. Another important aspect of AOM-SR is its application for desulfurization of wastewater where methane can be used as a sole electron donor. The use of methane for SR would close its cycle of utilization, decrease the emission of one of the most important greenhouse gases and reduce the risk of excess carbon source in the treated effluent.

For many years, methanotrophy was believed to be limited to oxic environments until the seventies of the last century when it was discovered during geochemical *in situ* studies in anaerobic marine sediments and waters (Martens and Berner, 1974; Barnes and Goldberg, 1976; Reeburgh, 1976), where methane was not accumulating before sulphate was exhausted. From the decrease of methane concentrations in the sulphate-reducing zone, it was concluded that methane must be consumed with sulphate (Figure 1.2). Studies have shown a stoichiometry according to equation 1.1.

$$CH_4 + SO_4^{2-} \rightarrow HCO_3^- + HS^- + H_2O \qquad\qquad \Delta G^{O'} = -16.6 kJ.mol^{-1} \qquad\qquad (1.1)$$

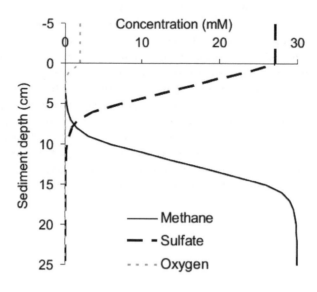

FIGURE 1.2 Typical methane, sulphate and oxygen concentration profiles in deep-sea AOM sediments where no convection takes place (Meulepas et al., 2010b).

AOM is mediated by a syntrophic consortium of methanotrophic archaea (ANME) and SRB. Three key groups of methanotrophic archaea, ANME-1, ANME-2 and ANME-3 have been identified (Hinrichs et al., 1999; Niemann et al., 2006). The archaeal group ANME-1 is distantly related to the *Methanosarcinales* and *Methanomicrobiales* (Hinrichs et al., 1999; Knittel et al., 2005) and appear as single cells, filaments or monospecies aggregates (Orphan et al., 2002; Valentine, 2002; Knittel et al., 2005). ANME-1 is dominant in diffusion driven methane seep (Niemann et al., 2005) and is an important organism in microbial mats of the Black Sea (Knittel et al., 2005). The biomass present in the most active cold seep is dominantly composed of aggregate methanotrophic consortia of ANME-2 or ANME-3 and SRB. In the aggregates, archaea and their partner SRB comprise up to 90% of the microbial biomass (Knittel et al., 2005). ANME-2 belongs to the order *Methanosarcinales* and is usually found in consortia with SRB belonging to the order *Desulfosarcina-Desulfococcus* (DSS) (Knittel et al., 2005). The recently discovered third group of anaerobic methanotrophs, ANME-3, is most closely related to *Methanococcoides* and *Methanolobus* and is typically found in aggregates with *Desulfobulbus*-like SRB (Niemann et al., 2006).

The genes required to perform all seven steps of methanogenesis from CO_2 were found present and actively expressed in ANME-2a. Thus, it is likely that AOM is carried out through a complete reversal of methanogenesis from CO_2 (Wang et al., 2014). A coupled two-step mechanism of AOM was proposed (Zehnder and Brock, 1979). In this mechanism, methane is first activated by methanogenic archaea, working in reverse, leading to the formation of intermediates, e.g., acetate or methanol. In a second step, the intermediates are oxidized to CO_2 coupled to sulphate reduction by other non-methanogenic members of the microbial community. The methane oxidation co-occurring with methanogenesis is called trace methane oxidation (TMO) and only a small portion of the methane is oxidized back to CO_2 (Meulepas et al., 2010a). No net methane oxidation by isolated methanogens has been reported so far. The main bottleneck of this process for biotechnological application is the extremely low growth rates of the responsible microorganisms (Meulepas et al., 2010b). To overcome this, full understanding of the process is needed. For example, understanding which compounds may be playing a role in the latter. Divergent results have been obtained when testing several compounds. Adding hydrogen, formate, acetate, methanol, carbon monoxide or methylamines reduced sulphate reduction rates in a sediment from Hydrate Ridge, suggesting that the SRB from this sediment were not adapted to those substrates (Nauhaus et al., 2002; Nauhaus et al., 2007). Similarly, Sørensen et al. (2001) excluded hydrogen, acetate and methanol as intermediates in the AOM-SR process saying that the maximum diffusion distances of the latter compounds, at *in situ* concentrations and rates were smaller than the thickness of two prokaryotic cell walls. Meulepas et al. (2010c) excluded acetate, formate, methanol, carbon monoxide and hydrogen, for an enrichment from Eckernförd Bay, as their concentration was 1000x higher than the concentrations at which no more Gibbs free energy can be conserved from their production for methane at the applied conditions.

Although hydrogen and formate were excluded as they could not be exchanged fast enough between syntrophic partners, shown by a process-based model, to sustain SR rates found by Nauhaus et al. (2007), it was shown that it can occur for acetate (Orcutt and Meile, 2008). Using a spherical diffusion-reaction model, hydrogen, formate and acetate were found to be thermodynamically and physically possible intermediates in AOM-SR (Alperin and Hoehler, 2010). A more recent study has discussed the possibility that AOM might not be an obligate syntrophic process but may be carried out by the ANME alone with zero-valent sulphur being an intermediate which is then disproportionated by the SRB (Milucka et al., 2012).This disparity in results is probably due to different groups of ANME and SRB and/or different environmental conditions. Thus, none of the mentioned compounds can be completely excluded as intermediate for AOM-SR.

1.5 POTENTIAL BIOTECHNOLOGICAL APPLICATION OF METHANE AS CARBON SOURCE FOR BIOLOGICAL SULPHATE REDUCTION

Research on anaerobic methane oxidation in marine environments has had its main focus on *in situ* conditions and unravelling the processes in marine sediments. Few published studies have evaluated the possible application of the AOM process coupled to SR. Since CH_4 is readily available and relatively cheap the direct use of methane for sulphate removal processes appears to be an excellent option. Methane which can be available as natural gas (80% CH_4) or biogas (50-75% CH_4) can be directly used as an electron donor for sulphate reduction. Four advantages have been identified in comparison to methods that use hydrogen as an electron source (Meulepas et al., 2010b):

1. The cost of the electron donor can be greatly reduced (Table 1.1)

2. The chemical catalysts used for steam reforming and the water-gas shift are easily polluted by hydrogen sulphide, present in the natural gas or biogas. Sulphide forms no problem when the CH_4 containing gas would be fed directly to the bioreactor.

3. The energy required to transfer the gas phase to the liquid phase is reduced as only one CH_4 can donate eight electrons, and one H_2 only two.

4. Substrate losses due to unwanted methanogenesis and acetogenesis (from hydrogen and CO_2) can be avoided, only microorganisms involved in AOM coupled to SR are able to grow in a methane-fed sulphate-reducing bioreactor.

TABLE 1.1 Prices and costs of electron donors for sulphate reduction (www.eia.gov).

Electron donor	Industrial market price (January 2014)	Required amount per kg of sulphate reduced	Electron donor cost [€.kg$_{sulphate}$$^{-1}$]
Ethanol	0.54 €.L^{-1}	0.40 L	0.216
Hydrogen	0.19 €.m^{-3}	0.934 m^3	0.18
Natural gas (80% CH_4)	0.11 €.m^{-3}	0.292 m^3	0.03

1.6 SCOPE AND ORGANIZATION OF THE THESIS

The main objectives of this research were to develop a process control strategy to optimize the input of a commonly used electron donor, i.e., lactate, and to study how to improve the feasibility of using methane as an electron donor for the biological sulphate reduction process treating wastewater. The specific objectives were: 1. To verify the suitability of a pS electrode for usage in process control; 2. To evaluate different input methods to control sulphide production; 3. To develop and calibrate a mathematical model capable of simulating the different processes taking place in SR process, i.e., accumulation of substrates; 4. To study how different co-substrates enhance or inhibit sulphate reduction coupled to anaerobic oxidation of methane performing *in vitro* incubations at high pressure.

The thesis is divided in 6 chapters. The present chapter gives a brief introduction on the content topics of the thesis. Chapter 2 discusses the different aspects to consider for the development of a bioprocess control. It discusses the mathematical models developed for sulphate reduction, the different sensors to be used and different control strategies. Chapter 3 presents an evaluation on how different input strategies will affect the control parameters (PID) of sulphide production in sulphate reducing bioreactors using a pS electrode. Chapter 4 presents how feast/famine conditions can affect accumulation of carbon and sulphur substrates in the biomass present in a sulphate reducing inversed fluidized bed reactor and describes a mathematical model developed to account for this accumulation process in such systems. Chapter 5 presents the effect of several labelled and non-labelled substrates added to the inoculum performing AOM-SR in high pressure vessels and low temperature (mimicking the original environment of the microbial consortia) on the activity rates of AOM and SR. A general discussion is given in Chapter 6 focusing on the implications of the findings of this thesis and recommendations for future research are given.

REFERENCES

Alperin MJ, Hoehler TM (2010) Anaerobic methane oxidation by archaea/sulfate-reducing bacteria aggregates: 1. Thermodynamic and physical constraints. Am J Sci 309: 869–957.

Barnes RO, Goldberg ED (1976) Methane production and consumption in anoxic marine sediments. Geology 4: 297–300.

Bijmans MFM (2008) Sulfate reduction under acidic conditions for selective metal recovery. Wageningen University, Wageningen, The Netherlands.

Cassidy J, Lubberding HJ, Esposito G, Keesman KJ, Lens PNL. 2015 Automated biological sulphate reduction: a review on mathematical models, monitoring and bioprocess control. FEMS Microbiol Rev. 39(6):823-53.

Dunn IJ, Heinzle E, Ingham J, Přenosil JE (2005) Automatic bioprocess control fundamentals. biological reaction engineering. pp. 161–179. Wiley-VCH Verlag GmbH & Co. KGaA.

Heinzle E, Dunn IJ, Ryhiner GB (1993) Modeling and control for anaerobic wastewater treatment. Adv Biochem Eng Biotechnol 48: 79–114.

Hinrichs KU, Hayes JM, Sylva SP, Brewer PG, DeLong EF (1999) Methane-consuming archaebacteria in marine sediments. Nature 398: 802–805.

Jagadeesh C, Sudhaker R (2010) Modeling, simulation and control of bioreactors process parameters – remote experimentation approach. Int J Comput Appl 1: 81–88.

Knittel K, Lo T, Boetius A, Kort R, Amann R (2005) Diversity and distribution of methanotrophic archaea at cold seeps. Appl Environ Microbiol 71: 467–479.

Koydon S (2004) Contribution of sulfate-reducing bacteria in soil to degradation and retention of COD and sulfate. Karlsruhe University, Germany.

Lens P, Vallero M, Esposito G, Zandvoort M (2002) Perspectives of sulfate reducing bioreactors in environmental biotechnology. Rev Environ Sci Biotechnol 1: 311–325.

Lens PNL, Visser A, Janssen AJH, Pol LWH, Lettinga G (1998) Biotechnological treatment of sulfate-rich wastewaters. Crit Rev Environ Sci Technol 28: 41–88.

Liamleam W, Annachhatre AP (2007) Electron donors for biological sulfate reduction. Biotechnol Adv 25: 452–463.

Martens CS, Berner RA (1974) Methane production in the interstitial waters of sulfate depleted marine sediments. Science 185: 1167–1169.

Meulepas RJW, Jagersma CG, Zhang Y, Petrillo M, Cai H, Buisman CJ, Stams AJ, Lens PNL (2010a) Trace methane oxidation and the methane dependency of sulfate reduction in anaerobic granular sludge. FEMS Microbiol Ecol 72: 261–271.

Meulepas RJW, Stams AJM, Lens PNL (2010b) Biotechnological aspects of sulfate reduction with methane as electron donor. Rev Environ Sci Bio/Technology 9: 59–78.

Meulepas RJW, Jagersma CG, Khadem AF, Stams AJM, Lens PNL (2010c) Effect of methanogenic substrates on anaerobic oxidation of methane and sulfate reduction by an anaerobic methanotrophic enrichment. Appl Microbiol Biotechnol 87: 1499–1506.

Milucka J, Ferdelman TG, Polerecky L, Franzke D, Wegener G, Schmid M, Lieberwirth I, Wagner M, Widdel F, Kuypers MMM (2012) Zero-valent sulphur is a key intermediate in marine methane oxidation. Nature 491: 541–546.

Nauhaus K, Boetius A, Krüger M, Widdel F (2002) *In vitro* demonstration of anaerobic oxidation of methane coupled to sulphate reduction in sediment from a marine gas hydrate area. Env Microbiol 4: 296–305.

Nauhaus K, Albrecht M, Elvert M, Boetius A, Widdel F (2007) *In vitro* cell growth of marine archaeal-bacterial consortia during anaerobic oxidation of methane with sulfate. Environ Microbiol 9: 187–196.

Niemann H, Elvert M, Hovland M, et al. (2005) Methane emission and consumption at a North Sea gas seep (Tommeliten area). Biogeosciences Discuss 2: 1197–1241.

Niemann H, Lösekann T, de Beer D, et al. (2006) Novel microbial communities of the Haakon Mosby mud volcano and their role as a methane sink. Nature 443: 854–858.

Orcutt B, Meile C (2008) Constraints on mechanisms and rates of anaerobic oxidation of methane by microbial consortia: process-based modeling of ANME-2 archaea and sulfate reducing bacteria interactions. Biogeosciences 5:1587–1599.

Orphan VJ, House CH, Hinrichs KU, McKeegan KD, DeLong EF (2002) Multiple archaeal groups mediate methane oxidation in anoxic cold seep sediments. Proc Natl Acad Sci U S A 99: 7663–7668.

Pind PF, Angelidaki I, Ahring BK, Stamatelatou K, Lyberatos G (2003) Monitoring and control of anaerobic reactors. Adv Biochem Eng Biotechnol 82: 135–182.

Postgate J (1984) The sulphate-reducing bacteria. 2nd ed. Cambridge University Press.

Reeburgh WS (1976) Methane consumption in Cariaco Trench waters and sediments. Earth Planet Sci Lett 28: 337–344.

Sawyer CN, McCarty PL, Parkin GF (2003) Chemistry for environmental engineering and science. McGraw-Hill.

Shin HS, Oh SE, Bae BU (1996) Competition between SRB and MPB according to temperature change in the anaerobic treatment of tannery wastes containing high sulfate. Environ Technol 17: 361–370.

Sørensen KB, Finster K, Ramsing NB (2001) Thermodynamic and kinetic requirements in anaerobic methane oxidizing consortia exclude hydrogen, acetate, and methanol as possible electron shuttles. Microb Ecol 42: 1–10.

Valentine DL (2002) Biogeochemistry and microbial ecology of methane oxidation in anoxic environments: A review. Antonie van Leeuwenhoek, Int J Gen Mol Microbiol 81: 271–282.

Villa-Gomez DK, Cassidy J, Keesman KJ, Sampaio R, Lens PNL (2014) Sulfide response analysis for sulfide control using a pS electrode in sulfate reducing bioreactors. Water Res 50: 48–58.

Wang FP, Zhang Y, Chen Y, He Y, Qi J, Hinrichs KU, Zhang XX, Xiao X, Boon N (2014) Methanotrophic archaea possessing diverging methane-oxidizing and electron-transporting pathways. ISME J 8: 1069–1078.

Zehnder AJ, Brock TD (1979) Methane formation and methane oxidation by methanogenic bacteria. J Bacteriol 137: 420–432.

2

Automated biological sulphate reduction: a review on mathematical models, monitoring and bioprocess control

CHAPTER 2

Automated
biological sulphate
reduction: a review
on mathematical
models, monitoring
and bioprocess
control

ABSTRACT

In the sulphate reducing process, bioprocess control can be used to regulate the competition between microbial groups, to optimize the input of the electron donor and/or to maximize or minimize the production of sulphide. As shown in this review, modelling and monitoring are important tools in the development and application of a bioprocess control strategy. Pre-eminent literature on modelling, monitoring and control of sulphate reducing processes is reviewed. This paper firstly reviews existing mathematical models for sulphate reduction, focusing on models for biofilms, microbial competition, inhibition and bioreactor dynamics. Secondly, a summary of process monitoring strategies is presented. Special attention is given to *in situ* sensors for sulphate, sulphide and electron donor concentrations as well as for biomass activity and composition. Finally, the state of the art of the bioprocess control strategies in biological sulphate reduction processes is overviewed.

This chapter has been published as:
Cassidy J, Lubberding HJ, Esposito G, Keesman KJ, Lens PNL. (2015) Automated biological sulphate reduction: a review on mathematical models, monitoring and bioprocess control. FEMS Microbiol Rev. 39(6):823-53

2.1 INTRODUCTION

2.1.1 IMPORTANCE OF BIOLOGICAL SULPHATE REDUCTION

Sulphate and other sulphur compounds are present in fresh water from geological origin or from the release of industrial activities. The production of edible oil, tannery, food processing, fermentation, coal mining and paper/pulp processing are industrial activities that emit elevated concentrations of sulphur compounds (Shin et al., 1996). In addition, elevated sulphate concentrations in fresh water bodies can be caused by seawater intrusion. In the absence of oxygen and nitrate, sulphate reduction by sulphate reducing microorganisms causes an increase in hydrogen sulphide concentration, which is toxic and causes an unpleasant smell and corrosion problems (Sawyer et al., 2003). Hydrogen sulphide is fatally toxic to humans, causing death within 30 minutes at gaseous concentrations of 800-1000 mg/L, and instant death at higher concentrations (Speece 1996). Therefore, it is important to desulphurize industrial wastewater prior to its discharge to the fresh water bodies. There are several methods for the removal of sulphur compounds from wastewater, including membrane filtration and chemical methods, which are expensive, and require a post-treatment of the brine. For high-strength sulphate containing wastewaters, biological sulphate removal is a cost-effective alternative (Lens et al., 1998).

Biological sulphate reduction is performed by a group of anaerobic bacteria, called Sulphate Reducing Bacteria (SRB). These bacteria are classified into two subgroups: autotrophic and heterotrophic SRB. Heterotrophic SRB (HSRB) use organic matter as the substrate, whereas autotrophic SRB (ASRB) use CO_2 as carbon source and H_2 as an electron donor (Liamleam and Annachhatre, 2007). Biological, anaerobic reduction of sulphate has been successfully applied for the treatment of sulphate contaminated wastewater from industries on a larger scale for many years, as it offers the possibility of an efficient treatment with low operation costs using various organic and easily utilizable carbon sources (Liamleam and Annachhatre, 2007). The end product is hydrogen sulphide (H_2S). Hence, this technological approach is very suitable for the treatment of waste streams containing dissolved metals. The metals can be precipitated simultaneously with the produced H_2S and removed as stable metal sulphide precipitates (Lewis, 2010).

Wastewaters from industries deficient in dissolved organic matter need to be supplemented with electron donors suitable for the sulphate reducing bacteria. Electron donors most commonly used are ethanol, lactate, formate, methanol, hydrogen, synthesis gas (80% H_2 and 20% CO_2), and CO. The application of the various electron donors for sulphate removal from various types of wastewater has been extensively reviewed (Liamleam and Annachhatre, 2007), with ethanol and hydrogen being the most commonly used electron donors in industrial applications. Lactate, in terms of energy and biomass yield, is reported as the best-suited carbon source (Postgate, 1984; Koydon, 2004), as many species of sulphate reducers can use it (Liamleam and Annachhatre, 2007). Acetate is a key intermediate in the breakdown of organic substances in anaerobic processes and has also been used as an electron donor in the sulphate reduction process (Lens et al., 2002). However, acetate is less suitable for high rate sulphate reduction processes, as some species of SRB cannot completely oxidize acetate and acetate utilization becomes the rate limiting step, even with excess sulphate levels.

Several bioreactor designs have been developed and applied successfully for biological sulphate reduction (Kaksonen and Puhakka 2007). These include batch reactors, sequencing batch reactors, continuously stirred tank reactors, anaerobic contact processes, anaerobic baffled reactors, anaerobic filters, fluidized bed reactors (up-flow and down-flow), gas lift reactors, up-flow anaerobic sludge blanket reactors, anaerobic hybrid reactors and membrane bioreactors.

2.1.2 BIOREACTOR MODELLING FOR SULPHATE REDUCTION PROCESSES

The performance of bioreactors can be evaluated either experimentally (empirical approach) or by simulation (modelling approach). The former is rather time and resource consuming due to the many experiments that are often needed and could possibly lead to irreversible modification of the biological processes under study. Making new experimental designs based on mathematical model simulations can reduce the number of required experiments to make predictions, whilst improving the effectiveness of the results. Hence, using a modelling approach to predict current and probable future events, whilst reducing the number of experiments, is an attractive way to get insight in the bioprocesses or to design bioreactors. The key problem when addressing bioreactor modelling is to find an appropriate model structure with reliable model parameters.

Modelling bioreactor performance typically starts from a well established theory of the processes that occur in the bioreactor. The bioreactor model is usually expressed in terms of non-linear differential equations, forcing a very detailed understanding of the processes going on in the bioreactor (Ryhiner et al., 1993, Dunn et al., 2005). Modelling is a crucial tool to identify the variables that significantly influence the system response and to give direction when establishing design criteria. In addition, a reactor model helps to identify possible causes for system malfunctioning or failure as well as in devising remedial measures (Kalyuzhnyi et al., 1998). Mathematical models, aimed at simulating the biochemical processes prevailing in the bioreactors, always need to be coupled to experimental studies in order to obtain calibrated and validated models to give decisive answers. Depending on the complexity and goals (see next section), the validated models can subsequently be used to: 1) address laboratory experimental procedures; 2) enhance the design and operation of the treatment systems; or 3) optimize the bioreactor process performance (Esposito et al., 2009).

2.1.3 BIOPROCESS CONTROL

Validated mathematical models are of great help for the development of advanced bioprocess control. In the sulphate reducing process, bioprocess control can be used to regulate the competition between microbial groups, to optimize the input of the electron donor and/or to maximize or minimize the production of sulphide. The latter is of great use for heavy metal recovery applications (Veeken et al., 2003). Bioprocess control also facilitates strategies for the management of the biocatalytic environment. In addition, control is necessary to induce the (micro-)organisms to produce substances in economically important amounts (Dunn et al., 2005). Anaerobic systems often show instabilities that may be caused either by toxic substances or overloading, which in turn may cause an irreversible collapse of the bioreactor. For this to be overcome, the process needs to be controlled to allow prolonged stable operation. Process optimization is closely linked with control. For example, the objectives of optimal control may be to maximize productivity, final concentration, yield, or to minimize effluent concentrations and energy costs (Dunn et al., 2003).

Significant progress has been made with the control of anaerobic systems, mostly methanogenic bioreactors (Pind et al., 2003). However, little research has been reported on the control of sulphate reducing bioreactors. Mathematical models reviewed in the sections below are the starting point for the development of such control strategies.

2.2 MODELS FOR BIOLOGICAL SULPHATE REDUCTION PROCESSES

Mathematical models are important tools in the understanding and optimization of the performance of biotechnological processes. The Monod model is commonly used to describe the kinetics of bacterial metabolism. The Monod model has been widely accepted, and offers mathematical simplicity. In this model, the bacterial growth rate (μ_j) is related to the concentration of the limiting substrate (S_i):

$$\mu_j = \hat{\mu}_j \, \frac{S_i}{K_{s,i} + S_i} \tag{2.1}$$

where, $\hat{\mu}_j$ is the maximum specific growth rate for biomass j and $K_{s,i}$ is the affinity constant of biomass j with respect to substrate i.

The Contois model is another very common model to describe microbial cell growth and substrate uptake kinetics. In this model, $K_{s,i}$ is considered to be dependent on the biomass concentration (X_j), thus:

$$\mu_j = \hat{\mu}_j \, \frac{S_i}{K_{s,i} X_j + S_i} \tag{2.2}$$

The maintenance energy requirement to explain the often observed decrease of yield at relatively low growth rates can be described by Pirt's equation:

$$\frac{v_i}{X_j} = \mu_j \frac{1}{Y_{X/S}} + m_s \tag{2.3}$$

where, v_i is the substrate utilization rate, m_s is the maintenance coefficient, $Y_{X/S}$ is the bacterial yield coefficient and μ_j is given by equations (2.1) and (2.2) or other kinetic models. The specific growth rate models described are used accordingly in the models reviewed below.

Comprehensive mathematical modelling of anaerobic processes is rather complex, as it involves complex dynamics of biological, chemical and physical subsystems with many interconnections between them. This section reviews models for biological sulphate reduction found in the literature. To help guide the readers selecting the most appropriate model, a summary of the main characteristics of the models discussed below can be found in Tables 2.1-2.4. Four families of models have been distinguished, describing: (a) anaerobic biofilms and granules; (b) microbial competition; (c) inhibition; and (d) bioreactor dynamics.

2.2.1 ANAEROBIC BIOFILM AND GRANULE MODELS

Biological sulphate reduction in anaerobic fixed growth reactors has been investigated extensively at lab-scale. In particular, it was pointed out that the composition of the microbial community influences the performance and stability of the overall biological sulphate reducing process (Celis et al., 2008). Modelling biofilms can help to further understand the dynamics of the microbial community, mass transport of substrates and their microbial conversion in the biofilm.

Mass transfer limitation of sulphate in UASB granules was studied theoretically by calculating the steady-state sulphate micro-profiles using a reference set of parameters obtained from experimental work (Overmeire et al., 1994). The model calculations showed that sulphate reduction can be limited in the UASB granules by mass transfer of sulphate into the granule (Figure 2.1).

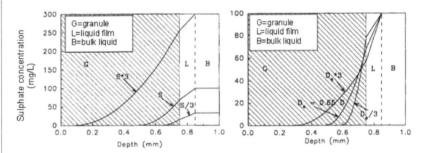

FIGURE 2.1 Influence of the bulk sulphate concentration, S, (left panel) and the effective diffusion coefficient, D_e, (right panel) on steady sulphate concentration profiles in a granule (Overmeire et al., 1994)

The parameters that mostly affected the diffusion of substrate in the granules were the sulphate concentration, the maximum sulphate utilization rate, the granular size and the effective diffusion coefficient. To reach these conclusions, the authors proposed a second-order differential equation that expresses the steady-state mass balance for sulphate ($S_{fSO_4^{2-}}$) on an elementary shell of volume within a spherical granule and includes Fick's law and Monod kinetics:

$$\frac{1}{L^2}\frac{d}{dr}\left(D_e L^2 \frac{dS_{fSO_4^{2-}}}{dr}\right) = \hat{v}_{SO_4^{2-}} \frac{S_{fSO_4^{2-}}}{K_{s,i} + S_{fSO_4^{2-}}} X_{SRB} \qquad (2.4)$$

with the following Newmann-type boundary conditions:

at L=0,

$$\frac{dS_{fSO_4^{2-}}}{dL} = 0 \qquad (2.5)$$

at L=L$_R$

$$D_e \left(\frac{dS_{fSO_4^{2-}}}{dL} \right) = k_l (S_{SO_4^{2-}} - S_{fSO_4^{2-}}) \qquad (2.6)$$

where, L is the distance normal to the granular surface with L=0 the centre of the granule and L$_R$ the radius; D$_e$ is the effective diffusion coefficient of the sulphate for transport in the granule; $\hat{v}_{SO_4^{2-}}$ is the maximum specific sulphate utilization rate (kgSO$_4^{2-}$.kg^{-1}.VSS.s^{-1}); k$_l$ is the mass transport coefficient of sulphate in the stagnant liquid film (m$_f^3$.m$_{gr}^{-2}$.s^{-1}) (f-fluid and gr-granule).

A mathematical model that incorporates the mechanisms of diffusion mass transport and Monod kinetics, similar to equations (2.4-2.6), but now under non-steady state conditions was developed for an anaerobic fixed biofilm reactor for phenol and sulphate removal (Lin et al., 2001). The model was validated with data from a pilot-scale column reactor. Batch tests were also conducted with the goal of determining the biokinetic coefficients used in the model. The model predictions agreed well for the non-steady state, but were not so successful under steady state conditions. Most likely, the authors did not include the effect of higher shear loss in thicker biofilms, which would have resulted in a higher suspended biomass concentration and therefore increasing total COD effluent concentrations. A sensitivity analysis of this process was performed, which showed that operational parameters such as, the hydraulic retention time and initial phenol concentration have a strong effect on the process efficiency (Lin et al., 2011). In addition, the results showed sensitivity to kinetic parameters (yield coefficient of phenol utilizing bacteria, Monod maximum specific utilization rate of phenol and phenol utilizing bacteria decay rate) and biofilm parameters (biofilm density of phenol utilizing bacteria and initial biofilm thickness).

A one-dimensional, multispecies biofilm model, which includes dual-substrate Monod kinetics and which describes the coexistence of denitrifiers and sulphate reducers in a H$_2$ fed membrane biofilm reactor is described by Tang et al. (2012). The model was calibrated and validated with experimental chemical and biological data. The authors assumed a steady state mass balance at any point in the biofilm and used Fick's law to describe the diffusivity. The model predicted that the onset of sulphate reduction occurred only when the nitrate concentration at the fibre's outer surface was low enough, leading to an equal growth rate for denitrifiers and sulphate reducers, i.e., the lower the concentration of nitrate the higher the SRB activity.

The biological, chemical and physical processes occurring in a sulphate reducing biofilm under dynamic conditions in an anaerobic fixed growth reactor were theoretically evaluated in the model of D'Acunto et al. (2011). A convection-diffusion model was developed.

The convection term governs the biofilm growth and the diffusion term was used for the substrate gradient throughout the biofilm. Three microbial groups were taken into account in this model: complete oxidizing SRB (SRB[(c)]), incomplete oxidizing SRB (SRB[(I)]) and acetate degraders (AcD) and three reaction components were considered (substrates and products): sulphate, lactate and acetate. The model was applied to simulate the effect of different COD/SO_4^{2-} ratios (Figure 2.2) and to predict the reactor's performance with respect to the volume fraction of bacterial species and substrate diffusion trends in the biofilm as a function of time.

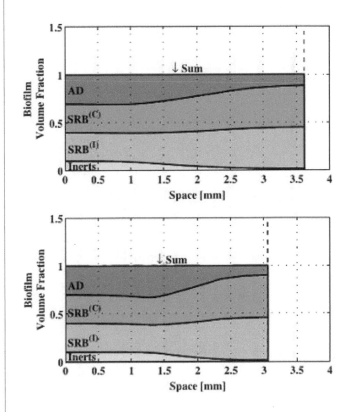

FIGURE 2.2 Effect of the COD/SO_4^{2-} ratio on the volumetric fraction of the bacterial species in the biofilm. Top: COD/SO_4^{2-} = 1, Bottom: COD/SO_4^{2-} = 0.25 (D'Acunto et al., 2011)

TABLE 2.1 Characteristics of selected anaerobic biofilm and granules models.

Goal	Substrate	Sludge feed	Microbial groups	Processes Involved	Hydrodyna mics	Inhibition	Bioreactor	Calibrated	Validated	References
Mass transfer limitation in UASB granules	Sulphate	UASB granules	SRB MA	Diffusive mass transport	Steady state	x	-	$\sqrt{a}$	x	Overmeire et al. (1994)
Phenol degradation with sulphate reduction in an anaerobic biofilm	Acetate	Anaerobic digested sludge	PUB SRB	Diffusive mass transport and Monod kinetics	Non-steady state	x	Fixed biofilm column	$\sqrt{}$	$\sqrt{}$	Lin et al. (2001) Lin et al. (2011)
Coexistence of denitrifiers and sulphate reducers	$NO_3^- - H_2$ $SO_4^{2-} - H_2$	-	DB SRB HB IB	Fick's law and dual-substrate Monod kinetic	Steady state	x	Membrane biofilm	$\sqrt{}$	x	Tang et al. (2012)
Population dynamics competition; Substrate dynamics	Lactate	-	SRB AcD	Diffusive and convective mass transport	Non-steady state	x	-	$\sqrt{a}$	x	D'Acunto et al. (2011)

a. calibrated with parameter values obtained from literature; √. Positive; x – negative; – information not available;

PUB- phenol utilizing bacteria; DB- denitrifying bacteria; HB- heterotrophic bacteria; IB- inert biomass

2.2.2 MICROBIAL COMPETITION

Sulphate reduction and methanogenesis are both involved in the final step of the degradation of organic matter in anaerobic environments (Figure 2.3).

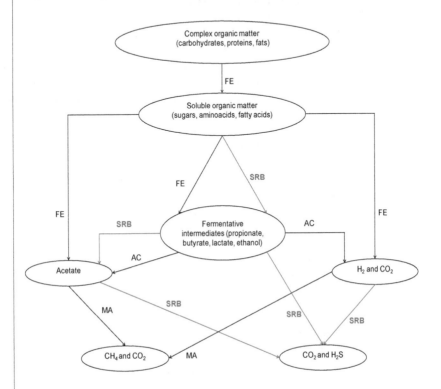

FIGURE 2.3 Pathways in the anaerobic degradation of organic matter under methanogenic (black) and sulphate reducing (red) conditions (adapted from Kalyuzhnyi et al. (1998)). (FE- fermenters, AC-acetogens, SRB-sulphate reducing bacteria, MA-methanogens).

Sulphate reduction and methanogenesis are both involved in the final step of the degradation of organic matter in anaerobic environments (Figure 2.3). Several microbial groups, fermenters (FE), acetogens (AC) and methanogenic archaea (MA) can use the same substrates as SRB, and therefore compete for it. In the anaerobic digestion process, SRB can compete with AC for volatile fatty acids and ethanol or with MA for acetate and hydrogen. Several factors can affect the outcome of this competition: COD/SO_4^{2-} ratio, type of seed sludge, sludge retention time, hydrogen sulphide inhibition, pH and nutrient limitation (Lens et al., 1998). This section overviews several models describing this competition.

Vavilin et al. (1994) simulated anaerobic degradation of organic matter by using a previously developed model that described the self-oscillating coexistence of MA and SRB (Vavilin et al., 1993). They calibrated and validated the model on the experimental data of Parkin et al. (1990), where anaerobic chemostats were maintained at changing acetate/sulphate influent concentrations. The authors concluded that both methanogenesis and sulphate reduction ceased when the COD/SO$_4^{2-}$ ratio was below 10/1. In this model, the Monod function (equation 2.1) was modified, adding two terms considering the pH and the sulphide concentration as inhibitory factors of the sulphate reducing process. The model simulations showed that pH and free hydrogen sulphide were the main factors for the system failure. The H$_2$S concentration acts as a trigger stimulating the positive feed-back loop between an increase in acetate and sulphate concentrations and a decrease in the pH level through the activity of the SRB and MA. Interestingly, simulations showed this feed back loop induced an oscillating coexistence between the two microbial groups (Figure 2.4). The modified function can be simplified if only one of the inhibitors has an effect on the process. Torner-Morales and Buitrón (2010) considered only pH to cause an inhibitory effect and thus, simplified the model by excluding sulphide inhibition from the equation. After calibration and validation, the model resulted in a predominance of incomplete oxidation of lactate over its complete oxidation.

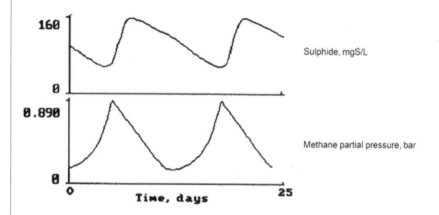

FIGURE 2.4 Self-oscillating coexistence of MA and SRB dependence on COD/SO$_4^{2-}$ ratio in anaerobic chemostats using acetate as electron donor (Vavilin et al., 1994).

Fomichev and Vavilin (1997) simplified the model of Vavilin et al. (1994) and created a reduced model of self-oscillating dynamics in an anaerobic system with sulphate reduction. Again, the authors calibrated the model with the experiments of Parkin et al. (1990). This reduced model is based on the competition between two microbial groups, MA and SRB, for the same substrate (acetate), product inhibition and pH influence on this inhibition. The authors concluded that, using this model, the oscillating phenomenon is due to hydrogen sulphide inhibition to both MA and SRB biomass growth and the influence of the pH on the equilibrium between ionized and non-ionized sulphide.

Similarly, the competition for acetate was studied by dynamic simulations of acetate utilizing SRB and MA in a UASB reactor treating volatile fatty acids (Omil et al., 1998). The simulations confirmed the long term competition between the acetotrophs. The main factors affecting the time required for acetate utilizing SRB to outcompete MA were pH (Figure 2.5), SRT and the size of the SRB population in the inoculum, under the assumption of a completely mixed high rate anaerobic reactor, with the sludge retention time (SRT) independent from the hydraulic retention time (HRT).

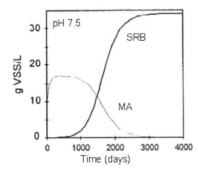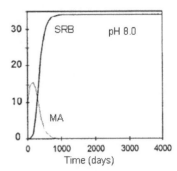

FIGURE 2.5 Effect of pH on the evolution of the SRB and MA population (Omil et al., 1998)

When lactate is used as electron donor, the competition between lactate oxidizers, such as SRB, and lactate fermenters (Figure 2.3) must be addressed in order to optimize its dosage. Kinetic properties of these pathways were determined and used to simulate the competition among the microbial species involved in anaerobic lactate degradation (Oyekola et al., 2012). The model was calibrated and validated on experimental data of laboratory scale chemostat cultures at different residence times and sulphate concentrations. The kinetic constants for lactate fermentation and lactate oxidation were calculated using the Monod (equation 2.1), Contois (equation 2.2) and Chen and Hashimoto kinetic expressions. The model also included the relationship between the kinetics of bacterial growth and lactate utilization rate (r_L) described by the Pirt equation (equation 2.3). On the basis of these simulations, the authors concluded that lactate oxidizers compete more efficiently with lactate fermenters for the carbon source at lactate concentrations below 5 g.L^{-1} and sulphide concentrations above 0.5 g.L^{-1}.

Using a dual-substrate Monod kinetics and previously described numerical method (D'Acunto et al., 2011), a model was developed to assess the microbial coexistence and competition between SRB[(c)], SRB[(l)], AC and MA growing on lactate (Mattei et al., 2014). The latter was simulated at various COD/SO$_4^{2-}$ ratios and showed that, although SRB are the most abundant throughout the biofilm for all simulations, the AC seem to occupy a greater area of the biofilm at higher COD/SO$_4^{2-}$ ratios since a higher COD load results in higher lactate concentrations throughout the biofilm thickness.

A simple model that can describe the competition between MA and SRB in a bioreactor fed with methanol was developed and calibrated on experimental data (Spanjers et al., 2002). The model was based on growth kinetics of hydrogen-consuming MA and SRB and methanol-oxidising AC. It describes three processes: growth of AC, SRB and MA, and four state variables: the methanol, hydrogen, sulphate and methane concentration. The conversion rates are assumed to be a function of the substrate concentration according to Monod kinetics (equation 2.1). However, the model did not give a good fit between the simulated and experimental values for the methane production rate. The authors hypothesized that acetate formation from methanol is an important step in the overall process and should have been included in the model.

When the substrate concentration is below a certain value, i.e., threshold substrate concentration, no substrate consumption occurs, even after a long incubation period. The following modification of the Monod kinetic expression (equation 2.1) that also takes into account the substrate threshold concentration (S_t) observed in bacterial growth has been proposed (Ribes et al., 2004):

$$\mu(S)=\mu_{max} \; \frac{S_i - S_t \, f}{K_s + S_i - S_t \, f} \qquad (2.7)$$

where, f and F are two sigmoid functions. f was added to avoid negative values of the specific growth rate when S_i is below but very close to S_t and F has still a certain non-zero value. F was added to smoothly lead the Monod function to zero at a certain value of S_i below S_t.

This kinetic expression was applied to an anaerobic model (Ribes et al., 2004) to simulate the competition for H_2 between MA and SRB in a thermophilic methanol-fed bioreactor. Using this model, the mathematical instabilities around the substrate threshold concentration that were observed when using the conventional kinetic model were avoided and thus, more acurate estimations of the biological process behaviour at very low substrate concentrations were obtained.

Wastewaters often contain multiple substrates. Thus, it is important to understand their effect on the competition between SRB and MA. The latter was included in a model of sulphate fed ideally mixed anaerobic reactors developed by Kalyuzhnyi and Fedorovich (1998). Four substrates were considered: sucrose, propionate, acetate and sulphate. The model was calibrated with data from laboratory studies. Subsequently, it was used to determine the effect of several factors on the outcome of the competition. The model was able to describe the steady state performance of the reactor and the increase of total COD converted by the SRB relative to that converted by the MA under the different HRTs imposed. The specific growth rates were determined with the same equations of Kalyuzhnyi and Fedorovich (1997). The material balances, however, were slightly different because of the different hydrodynamics. The gas volumetric flow rate (G) from the reactor and the mass transfer rate to the gas phase (M_i) are given by:

$$G = \sum_i (M_i V_g)V \qquad\qquad M_i = k_L a \left(S_i^* - \frac{p_i}{K_H} \right) \qquad (2.8)$$

where V_g is the specific volume of gas; $k_L a$ is the mass transfer coefficient; p_i is the partial pressure; K_H is the Henry´s constant; * stands for undissociated.

The material balance for the influent substrates (sucrose, propionate, acetate and sulphate) can be written as:

$$\frac{d}{d_t} S_i = D(S^0 - S_i) + v_i \qquad (2.9)$$

where D is the dilution rate (HRT^{-1}).

The behaviour of each bacterial group in the reactor can be described as:

$$\frac{d}{d_t} X_j = \mu_j X_j - \frac{(1 - ER_j)X_j}{HRT} - bX_j \qquad (2.10)$$

where b is the decay coefficient; ER_j is the efficiency of retention of bacterial group j.

Frunzo et al. (2012) presented a mathematical model that was able to simulate the biological, chemical and physical processes prevailing in a biological sulphate reduction gas-lift reactor under dynamic conditions. The model considers the kinetics of microbial growth and decay. In particular, the model takes five groups of bacteria into account, i.e., HSRB, ASRB, homoacetogenic bacteria (HB), MA, and AcD; and six components (substrates and products), i.e., H_2, SO_4^{2-}, CO_2, acetate, H_2S and inert material. The mass balance equations for both liquid and gas phases and both substrates and bacterial groups were:

$$\frac{d[S]_{(l)}}{d_t} = \frac{Q}{V} ([S]_{(l),in} - [S]_{(l)}) - k_L a([S]_{(l)} - [S^*]_{(l)}) + \sum_{j=1...N} \rho_j v_j \qquad (2.11)$$

$$\frac{d[S]_{(g)}}{d_t} = \frac{V}{V_g} k_L a([S]_{(l)} - [S^*]_{(l)}) - \frac{Q_g}{V_g} [S]_{(g)} \qquad (2.12)$$

$$\frac{dX}{dt} = \frac{Q}{V} [R(\beta-1)-1][X] + \sum_{j=1...N} \rho_j v_j \qquad (2.13)$$

where, $[S]_{(l),in}$ and $[S]_{(l)}$ are the molar concentration of the specific gas in the digester influent and the aqueous phase, respectively; Q_g is the total effluent gas flow rate; V_g is the gas phase volume; β is the thickening ratio in the sedimentation tank and R is the ratio between the sludge recycle flow rate and the influent flow rate.

The model was calibrated and validated using an experimental study (Esposito et al., 2003). It adequately simulates the bioconversion processes and predicts properly the effects of the variations of the operational conditions on the bacterial competition in the gas-lift reactor, which can be subsequently used for process optimization and control.

TABLE 2.2 Characteristics of selected microbial competition models.

Goal	Substrate	Sludge feed	Microbial groups	Processes Involved	Hydrodynamics	Inhibition	Bioreactor	Calibrated	Validated	References
Self-oscillating coexistence	Acetate	-	SRB MA	Moser and Monod kinetics	Non-steady state	sulphide and pH	Chemostat	√a	√a	Vavilin et al. (1994)
Competition for substrate	Lactate	UASB granules	SRB[c] SRB[l]	Monod kinetics	Non-steady state	pH	SBR	√	√	Torner-Morales et al. (2010)
Self-oscillating coexistence	Acetate	-	SRB MA	Monod kinetics	Non-steady state		Chemostat	√a	√a	Fomichev and Vavilin (1997)
Dynamic competition for substrate	Volatile fatty acids	UASB granules	SRB MA	Monod kinetics	Non-steady state	x	UASB	√	x	Omil et al. (1998)
Competition for substrate	Lactate	facultative pond treating sewage	SRB LF	Pirt, Monod and Contois kinetics	Non-steady state	x	Chemostat	√	x	Oyekola et al. (2012)
Coexistence and competition in biofilm	Lactate		SRB[c] SRB[l] AC MA	Monod kinetics	Non-steady state	x	-	-	-	Mattei et al. (2014)
Competition for substrate	Methanol		SRB MA	Monod kinetics	Non-steady state	x	EGSB	√	x	Spanjers et al. (2002)
Competition for multiple substrate	Sucrose, propionate, acetate	Anaerobic granular sludge	SRB MA	Dual substrate Monod kinetics	Steady state	First order kinetic	UASB	√	√	Kalyuzhnyi and Fedorovich (1998)
Competition for substrate	Hydrogen	-	SRB MA	Modified Monod kinetics	-	x	EGSB	√	x	Ribes et al. (2004)
Effect of operational conditions on the bacterial competition	H_2 and CO_2	-	HSRB ASRB HB MA	Monod kinetics	Non-steady state	x	Gas lift	√	√	Frunzo et al. (2012)

a. calibrated with parameter values obtained from literature; √. Positive; x – negative; - information not available

2.2.3 INHIBITION

In addition to the structure of the microbial community and microbial competition, inhibition may have a strong effect on biochemical processes by decreasing the conversion or growth rate affecting the overall performance. Inhibition can be induced by substrate or product concentrations. Commonly, inhibition increases with an increase in the inhibitor concentration, leading to a gradual decrease in the specific substrate utilization rate. Therefore, it is important to understand the inhibition kinetics.

The sulphide product inhibition of *Desulfovibrio desulfuricans* in batch experiments was modelled (Okabe et al., 1995) using a simplified Monod model (equation 2.1), assuming $S_i >> K_{s,i}$. In the batch experiments, the inhibition coefficient, $K_{i,k}$ for the maximum specific growth rate was determined to be 251 mgS.L^{-1}. Since there was no significant change in lactate utilization rate below 437 mgS.L^{-1} and the only varying parameter was the g of cells produced, the authors were able to use the cell yield ($g_{cell}.g_{lactate}^{-1}$) to represent inhibition. In the chemostat experiments, at pH 7.0, the cell yield was halved at a sulphide concentration of approximately 250 mgS.L^{-1}, which was very close to the $K_{i,k}$ determined in the batch experiments. Thus, the non-competitive inhibition model adequately described sulphide inhibition of *D. desulfuricans* in the batch experiment. The authors showed that it is crucial to distinguish between sulphide inhibition of cell yield (growth) and the activity (lactate utilization rate). In their study, the calculated maintenance coefficient increased at total sulphide concentrations above 200 mgS.L^{-1}, thus leading to a decrease of the cell yield but not affecting lactate consumption.

When high biomass concentrations are present in the bioreactor, one might consider using a modified Contois model (Moosa et al., 2002):

$$\mu_j = \left(\hat{\mu}_j \frac{S}{K_{s,i} X + S} - b \right) \frac{X}{Y_{X/S}} \tag{2.14}$$

The latter was used by Moosa and Harrison (2006) to determine the microbial growth parameters under different pH conditions which altered the sulphide speciation in an acetate fed mixed population of complete oxidizing SRB treating acid mine drainage. The authors showed that the volumetric sulphate reduction rate and specific growth rate of SRB correlate inversely to the concentration of undissociated H_2S, i.e. at lower pH (<7.0). In fact, at pH 7.8, the inhibition on sulphate conversion was observed when the sulphide concentrations were above 750 mg.L^{-1}, which is higher than the value of 250 mg.L^{-1} observed in the work of Okabe et al., 1995. Thus, the speciation is a crucial factor when determining the inhibition.

Reis et al. (1990) proposed an empirical model, which involves the kinetics of acetate inhibition on lactate fed SRB in the absence of hydrogen sulphide. The authors concluded that the undissociated acid is inhibitory for the SRB at pH 5.8-7.0. The model gave a good fit to the experimental data and predicted a concentration for the undissociated acetic acid of 54 mg.L^{-1} (=K_{AcH}) that leads to a 50% inhibition. The same authors extended their work and proposed an empirical model to describe the concomitant inhibition of hydrogen sulphide and acetic acid on lactate fed sulphate reducing bacteria (Reis et al., 1992). The maximum concentration for hydrogen sulphide obtained from this equation was 547 mg.L^{-1}. The inhibition kinetics of hydrogen sulphide were described mathematically using a non-competitive inhibition model similar to (2.14). The study also showed that when there is high undissociated acetic acid concentration, the effect of hydrogen sulphide is not relevant. On the other hand, when there is high sulphide concentration, an increase in acetic acid concentrations leads to a significant decrease of the specific growth rate.

Biological sulphate reduction is an increasing popular method for the treatment of acid mine drainage and wastewaters from metal processing, mining and petrochemical industries which contain high concentrations of both heavy metals and sulphate. Thus, it is important to also study the inhibition of sulphate reducers by these heavy metals. A mathematical model for cadmium removal by precipitation with biogenic sulphides produced by a single bacterium species was developed taking into account the inhibition of hydrogen sulphide (López-Pérez et al., 2013). A modified Levenspiel inhibition model was used and a high correlation (0.99) was obtained between simulation and experimental results. It predicted inhibitory effect of Cd^{2+} for concentrations above 190 mg.L^{-1}:

$$\frac{dX}{dt} = v_i \left(1 - \frac{H_2S}{K_{H_2S}}\right)^\alpha \left[\frac{S_{SO_4^{2-}}}{k_{s,SO_4^{2-}} + S_{SO_4^{2-}}}\right] \left[\frac{S_{Cd^{2+}}}{K_{Cd} + S_{Cd^{2+}}}\right]^\eta X\, S_{Lactate}^{\varepsilon} - bX S_{Lactate}^{\varepsilon} \qquad (2.15)$$

where, k_{Cd} is the inhibition constant for Cd^{2+}; α is the exponential term for Luong model; η is the exponential term for Moser model; and ε is the exponential term for the lactate concentration.

Gonzalez-Silva et al. (2009) studied the inhibition of the specific substrate utilization rate of ethanol fed anaerobic granular sludge by iron, cadmium and sulphide using batch tests. For this purpose, the authors used the Monod equation (equation 2.1) to determine the kinetic parameters (v_L and Ks) and fitted the Andrews-Noack (Andrews, 1968) non-competitive inhibition model (equation 2.16) to calculate the inhibition constant, K_I. At pH 6.2-6.6, the IC$_{50,totalsulphide}$ and IC$_{50,H_2S}$ were 397 and 291 mgS.L^{-1}, respectively. These results again support the importance to take into account the chemical speciation of H$_2$S. For Fe^{2+} and Cd^{2+}, the inhibition occurred at concentrations above 467 and 1012 mg.L^{-1}, respectively. The Andrews-Noack non-competitive inhibition model for substrate inhibition can be considered as a multiplicative Monod model (Andrews, 1968):

$$\mu_j = \hat{\mu}_j \frac{S_i}{(K_{s,i} + S_i)} \frac{K_{I,k}}{K_{I,k} + I_k} \qquad (2.16)$$

where I_k is the inhibitor concentration.

Most wastewaters contain nitrate in adition to sulphate. It is thus important to understand how to optimize the process of simultaneous or sequential reduction of both terminal electron acceptors. To understand the effects of one another, Xu et al. (2014) developed a model to simulate the co-reduction of nitrate and sulphate. For this model, the authors used Monod kinetics (equation 2.1) incorporated with a competitive inhibition modifier to predict the effects of the anions on the nitrate and sulphate reduction rates. Although the authors verified simultaneous removal of nitrate and sulphate, the sulphate reduction rate was retarded with 56% in the presence of nitrate.

TABLE 2.3 Characteristics of selected inhibition models.

Goal	Substrate	Sludge feed	Microbial groups	Processes Involved	Hydrodynamics	Inhibition	Bioreactor	Calibrated	Validated	References
Sulphide inhibition	Lactate	Desulfovibrio desulfuricans	Desulfovibrio desulfuricans	Monod kinetics	Steady state	Non-competitive	Batch/Chemostat	√	x	Okabe et al. (1995)
pH effects	Acetate	SRB enriched mixed culture	SRB	Modified Contois kinetics	Non-steady	Non-competitive	Chemostat	√	x	Moosa and Harrison (2006)
Acetate inhibition	Lactate	Anaerobic digested sludge	SRB	Polynomial fit	-	Non-competitive	Chemostat	√	√	Reis et al. (1990)
Acetate and sulphide inhibition	Lactate	Anaerobic digested sludge	SRB	Monod kinetics	-	Non-competitive / Competitive	Chemostat	√	√	Reis et al. (1992)
Cadmiun inhibition	Lactate	Desulfovibrio alaskesnis 6SR	Desulfovibrio alaskesnis 6SR	Levenspiel inhibition model	Non-steady	Non-competitive	Batch	√	√	López-Pérez et al. (2013)
Iron, cadmium and sulphide inhibition	Ethanol	Anaerobic granular sludge	SRB	Monod kinetics	Pseudo-steady state	Non-competitive	UASB	√	x	Gonzalez-Silva et al. (2009)
Nitrate inhibition	Lactate	Mixed micobrial biomass	-	Monod kinetics	Non-steady	Competitive	Batch	√	x	Xu et al. (2014)

√. Positive; x – negative; - information not available

2.2.4 BIOREACTOR DYNAMICS

The models described so far aimed at predicting or stimulating the microbial performance. Modelling can also be used as a tool to develop scaling-up criteria, i.e., if the substrate removal is established as the model input, design models give the reactor size as an output. This is of great importance when designing a full-scale reactor. In this section, such models are discussed based on the distinction between (i) continuous stirred-tank and (ii) plug flow reactors.

2.2.4.1 CONTINUOUS STIRRED-TANK REACTORS

Gupta et al. (1994a) developed the first design model for sulphate fed anaerobic reactors. The model described the complex chemistry involved in anaerobic digestion of organic matter incorporating the various buffer systems, acid-base and liquid-gas equilibria (carbon dioxide, hydrogen sulphide, ammonia, methane, nitrogen and water vapour), ionic interactions and metal precipitation. The overall mass balance equations of the various components include liquid as well as gas phase concentrations in order to accurately predict the effluent gas production rate and composition. The model was calibrated and validated with previous experiments where three different substrates were used (acetic acid, methanol and formic acid) and operating under two different conditions (methanogenic and sulphate-reducing). Iron was added to precipitate the sulphide produced (Gupta et al., 1994b).

The model was able to predict the reactor performance fairly well for both steady state and batch experiments under ideally mixed conditions (Figure 2.6). The general mass balance for a specific component in a CSTR is given by:

$$\text{net rate of accumulation} = \text{rate}_{in} - \text{rate}_{out} \pm \text{rate of reaction} \qquad (2.17)$$

The mass balance equation for the biomass assuming constant volume of the reactor's liquid was also incorporated into the model to calculate the amount of substrate converted into biomass. This equation is needed to close the substrate mass balance:

$$V \frac{dX_j}{dt} = Q (X_{j,in} - X_j) + V (Y_{X/S} v_i X_j - b X_j) \qquad (2.18)$$

where V is the liquid phase volume and Q is the influent flow rate.

A more complex structure of the biological subsystem for the description of the dynamic and steady state behaviour of an anaerobic digester for the treatment of high strength sulphate wastewaters was developed by Knobel and Lewis (2002). The model, applicable for a number of carbon sources (both simple and complex) and for different microbial groups accounted for inhibition by pH, sulphide, hydrogen and fatty acids and was valid for a number of reactor types. A first order model was used to describe the hydrolysis rate and the Monod model (equation 2.1) was used for the specific biomass growth rate.

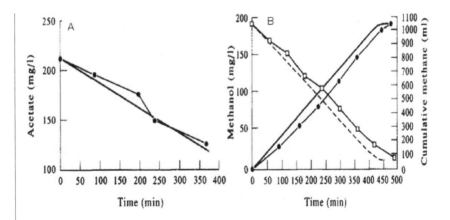

FIGURE 2.6 (A) Comparison of the model simulation with results from the batch spike experiments with acetate as electron donor (•experimental - model simulation). (B) Comparison of the model simulation with results from the batch spike experiments with methanol as electron donor (□ experimental methanol, -- model simulation methanol; •experimental methane, - model simulation methane) (Gupta et al., 1994a).

Competitive, non-competitive and uncompetitive inhibition models were taken into consideration for unionized fatty acids or sulphide. The inhibitory effects of a too high or too low pH were also accounted for. The model was shown to be capable of predicting a number of different scenarios, including the time dependent sulphate and COD concentrations in molasses fed packed bed and UASB reactors (Figure 2.7). Also the dynamic sulphate conversion rate in a gas lift reactor fed with hydrogen and carbon dioxide was well fitted. The calibration of the model was done with data from the literature.

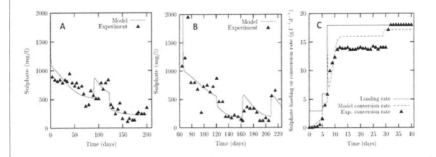

FIGURE 2.7 (A) Actual and simulated sulphate concentrations in a molasses fed packed bed reactor (A), UASB (B) and actual simulated sulphate loading and conversion rates in a hydrogen fed gas lift reactor (C) (Knobel and Lewis, 2002).

The Anaerobic Digestion Model No.1 (ADM1) (Batstone et al., 2002) is a structured model that includes disintegration and hydrolysis, acidogenesis, acetogenesis and methanogenesis as the steps in anaerobic biodegradation. Additional blocks describing the sulphate reduction processes were later included by Fedorovich et al. (2003). The revised model was applied to describe a long-term experiment on sulphate reduction with volatile fatty acids as the substrate in an upflow anaerobic sludge bed reactor and was able to predict the outcome of the competition among AC, MA, and SRB for these substrates. The model was validated on an experimental study which considered different operating regimes of granular sludge bed reactors with effluent recycle. The kinetics of sulphate reduction processes were introduced following the principles of ADM1 taking into account both the electron donor S_S (organic substrate or hydrogen) and electron acceptor S_k (SO_4^{2-}) concentration in an extended Monod model:

$$\mu_j = \hat{\mu}_j \frac{S_s}{(K_{s,s} + S_s)} \frac{S_k}{K_{s,k} + S_k} \qquad (2.19)$$

Similarly, Poinapen and Ekama (2010) extended an anaerobic digestion model (Sötemann et al., 2005) by adding the biological, chemical and physical processes associated to biological sulphate reduction. For this purpose, the authors considered Monod kinetics with concomitant inhibition by undissociated H_2S and pH. A more stable inhibition function (equation 2.20) was used instead of a first order inhibition function. The model was successfully validated with experimental results obtained from UASB reactors fed with several COD/SO_4^{2-} ratios at different temperatures.

$$\mu_j = \hat{\mu}_j \frac{S_{iF(pH)}}{(K_i + S_i)} \exp\left[-\left(\frac{[H_2S]}{0{,}60056K_{H_2S}} \right)^2 \right]\left(\frac{S_{SO_4^{2-}}}{k_s + S_{SO_4^{2-}}} \right) \qquad (2.20)$$

A model for sulphate reduction in a liquid-solid fluidized bed reactor was developed by Nagpal et al. (2000) with the aim to identify the limiting factors and design modifications, allowing enhancement of SRB growth. The model was calibrated and validated and showed a good match between the simulation results and experimental data using ethanol as electron donor. Two substrates (sulphate and ethanol) and two products (acetate and sulphide) were considered as well as the inhibition of the SRB growth rate by ethanol and acetic acid. The model took into consideration both the biomass attached to the beads and the biomass in the liquid phase. The model suggested that a significant increase in the sulphate reduction capacity of the system is possible by increasing the volume of the bed relative to the total liquid volume of the fluidized bed reactor.

Three steady state mathematical models for the design of H_2/CO_2 fed gas-lift reactors, aiming at sulphate reduction, were developed by Esposito et al. (2009). The proposed models gave the reactor volume required for an assigned sulphate removal efficiency. The simulations performed showed that the size of these reactors highly depends on the number and type of trophic groups present in the sludge. Thus, knowledge on the microbial groups present is crucial to obtain the required volume properly. Their model 1A takes into account 2 groups of heterogenic bacteria (HB and HSRB), three substrates (H_2, SO_4^{2-} and acetate) and two products (acetate and sulphide). Model 1B considers the same assumptions as Model 1A, but it is based on the hypothesis that besides HSRB and HB, also MA may grow on H_2/CO_2 with CH_4 as the end product; whereas acetoclastic methanogens do not grow in the reactor. Model 2 is based on the hypothesis that ASRB is the dominant microbial group in the reactor and considers the same assumptions as the previous models. Thus, model 2 takes into account one group of bacteria (ASRB), two substrates (H_2 and SO_4^{2-}) and one product (H_2S). The steady state design Model 1B (Esposito et al., 2009) was validated in the work of Frunzo et al. (2012) presented in the section 2.2.2.

2.2.4.2 PLUG FLOW REACTORS

In plug flow reactor configurations with axial dispersion and reactions, the spatial distribution of any component N in the liquid phase can be written by the following equation:

$$\frac{\partial}{\partial t} N(z,t) = \frac{\partial}{\partial z}\left[D_a(z,t) \frac{\partial}{\partial z} N(z,t) \right] - \frac{\partial}{\partial z}[W(z,t)N(z,t)] + q(z,t) - M(z,t) \qquad (2.21)$$

with the first term on the right hand side of the equation characterizing the degree of mixing by gas induced dispersion, D_a represents the axial dispersion coefficient, the second term of the equation determining a convective part of mass transfer in the flow direction, W represents the superficial velocity, the third and fourth elements represent the net biological production/consumption rate and transfer rate from the liquid to gas phase for the component N, respectively.

The behaviour of a bacterial group in a plug flow system can be described as:

$$\frac{\partial}{\partial t} X_j(z,t) = \frac{\partial}{\partial z}\left[D_j(z,t) \frac{\partial}{\partial z} X_j(z,t) \right] - \frac{\partial}{\partial z}[W_j(z,t)X_j(z,t)] + (\mu_j - k_d) X_j(z,t) \qquad (2.22)$$

The first attempt to develop a model for the concentration gradients on substrates, intermediates, products and bacteria in sulphate fed UASB reactors was undertaken by Kalyuzhnyi and Fedorovich (1997). The approach was generalized, which resulted in the development of the dispersed plug-flow model of sulphate fed UASB reactors (Kalyuzhnyi et al., 1998). The model was calibrated and validated with experimental studies of UASB reactors with acetate, propionate and sucrose as COD source. It adequately described the experimental data on the functioning of UASB reactors both during the start up with almost non-sulphate adapted seed sludge and during the stage when mature granular sulphidogenic sludge had been formed. It includes fermenters, AC, SRB and MA. Thus, the model could be used for maximization of the sulphide yield and model-based process control. The model includes four blocks: 1) kinetics, described by a non-competitive inhibition model; 2) physico-chemical parameters; 3) hydrodynamics and 4) mass balances for gas, soluble substrates and bacterial groups.

A model similar to Kalyuzhnyi et al. (1998) (equation 2.22) composed by 8 partial differential equations using single and dual-substrate Monod-type kinetics for biomass growth rate was developed to simulate the processes in a horizontal-flow anaerobic immobilized biomass (HAIB) bioreactor (Rodriguez et al., 2011). It considered that the concentrations of substrates and products were subjected to both the plug flow hydrodynamics and metabolic reactions. The model comprised AC, SRB[(C)] and SRB[(I)] as microbial groups and ethanol as the initial carbon source and presented good agreement with the data.

TABLE 2.4 Characteristics of selected bioreactor models.

Goal	Substrate	Sludge feed	Microbial groups	Processes Involved	Hydrodynamics	Inhibition	Bioreactor	Calibrated	Validated	References
Bioreactor model with simple biological system structure	Acetic acid, methanol, formic acid	Mixed culture methanogenic and sulphate reducing reactors	SRB MA	Monod kinetics	Steady-state and batch experiments	x	Chemostat	√	√	Gupta et al. (1994)
Bioreactor model with complex biological system structure	1.Molasses 2.H$_2$ and CO$_2$	-	2 FE 2 AC 2 MA 5 SRB	Monod kinetics	Steady and non-steady state	Non-competitive	1.Fed packed bed/UASB 2.Gas lift	√	√	Knobel and Lewis (2002)
Inclusion of sulphate reduction in the ADM1.	Volatile fatty acid	Anaerobic granular sludge	AC MA SRB	Monod kinetics	Non-steady state	First order kinetics	UASB	√	√	Fedorovich et al. (2003)
Define reactor's volume for given efficiency	H$_2$ and CO$_2$	-	1A.HB/HSRB 1B. HB/HSRB/MA 2.ASRB	Monod kinetics	Steady state	x	Gas lift	√	√a,*	Esposito et al. (2009)
Plug flow model	Acetate Propionate Butyrate Hydrogen	-	FE AC MA SRB	Monod kinetics and dual substrate Monod kinetics	Steady state	First order kinetics	UASB	√a	√	Kalyuzhnyi et al. (1998)
Plug flow model	Ethanol	UASB granules	SRBᶜ SRBᶠ AC	Dual substrate Monod kinetics	Steady state	x	HAIB	√	√	Rodriguez et al. (2011)
Inclusion of sulphate reduction in AD model	Primary sewage sludge	-	SRB	Monod kinetics	Steady state	Sulphide and pH	UASB	√	√	Poinapen and Ekama et al. (2010)
Performance enhancement	Ethanol	mixed culture of SRB	SRB	Dual substrate Monod kinetics	Non-steady state	Non-competitive	Liquid-solid fluidized bed	√	√	Nagpal et al. (2000)

*Model 1 A validated in the work of Frunzo et al. 2012

a. Calibrated with parameter values obtained from literature; √. Positive; x – negative; - information not available

2.2.5 KINETIC PARAMETERS

Not only the choice of a suitable mathematical model structure is important but also the choice of the kinetic parameters to be used in any mathematical model is of crucial importance to obtain accurate and valid results. Values of kinetic parameters are initially determined experimentally and can be better calibrated to obtain the best fit (Keesman, 2011). This calibration can be done using several performance measures (Janssen and Heuberger, 1995) which aim at finding the best value that minimizes the difference between experimental and simulated data. Tables 2.5-2.8 overview the kinetic values used in the previously described models. It should be noted that the coefficient of variation is significantly high for each group of kinetic parameters used in all models reported, even for the same substrate. This is probably due to a high variability in the experimental conditions under which the parameters were estimated. Thus, careful attention must be given when choosing the parameter values to be used.

TABLE 2.5 Kinetic parameters for SRB with acetate as a substrate.

Microbial groups	Substrate (i)	Ks (gSO$_4^{2-}$.L^{-1})	Kj (g.L^{-1})	Y (g biomass. g^{-1} substrate)	Ki (g.L^{-1})	μ (d^{-1})	ν (d^{-1})	Kd (d^{-1})	Kinetics	References
SRB	Acetate	0.012	0.13		K$_{H_2S}$=0.035 K$_i$=5.10/4.85 K$_{OH}$=8.0/8.3	0.64	-	-	Moser and Monod	Vavilin et al. (1994)
SRB	Acetate	-	0.00084	0.05±0.015	-	-	0.36[a]	-	Monod	Gupta et al. (1994)
SRB	Acetate	0.13	-	0.07	K$_{H_2S}$=0.036/0.07	-	8	0.03	Single Monod	Fomichev and Vavilin (1997)
SRB	Acetate	0.033	0.055	0.050	-	0.120	-	0.005	Single and dual Monod	Omil et al. (1998)
SRB	Acetate	0.0192	0.024	0.041	K$_{H_2S}$=0.285	0.51	-	0.025	Dual Monod	Kalyuzhnyi and Fedorovich (1998)
SRB	Acetate	0.0192	0.024	0.033	K$_{H_2S}$=0.164	0.612	-	0.0275	Single and dual Monod	Kalyuzhnyi et al. (1998) Poinapen and Ekama (2010)
SRB	Acetate	0.0096	0.22	0.0342	K$_{H_2S}$=0.26554	-	0.24282	0.015	Dual Monod	Fedorovich et al. (2003)
SRB	Acetate	-	0.131	-	-	1.416	-	0.816	Modified Contois	Moosa and Harrison (2006)
HSRB	Acetate	0.00045	0.015	0.12	-	4.9	-	0.04	Dual Monod	Esposito et al. (2009)
SRB	Acetate	0.01069	0.00975	0.0875	-	-	9.89[b]	0.0158	Dual Monod	Lin et al. (2001) Lin et al. (2011)
SRB[c]	Acetate	0.51	0.61	0.012	-	0.025	-	0.022		Rodriguez et al. (2011)
Cv		1.881	1.444	0.556		1.343		2.254		

a – expressed as g.L^{-1}.d^{-1}; b – expressed as g.g^{-1}VSS.d^{-1}; Cv – coefficient of variation; minimum and maximum in italic.

TABLE 2.6 Kinetic parameters for SRB with lactate as a substrate.

Microbial groups	Substrate (l)	Ks (g SO$_4^{2-}$.L^{-1})	KJ (g.L^{-1})	Y (g biomass. g^{-1} substrate)	Ki (g.L^{-1})	μ (d^{-1})	ν (d^{-1})	Kd (d^{-1})	Kinetics	References
SRB	Lactate	-	-	-	KAc=0.054	3.288	-	-	Polynomial fit	Reis et al. (1990)
SRB	Lactate	-	-	0.1489 (SO$_4^{2-}$)	KH=3.25x10^{-7} KOH=2.87x10^{-6} KAcH=0.054	8.064	-	-	Monod	Reis et al. (1992)
Desulfovibrio desulfuricans	Lactate	-	0.00235	0.036	KH$_2$S=0.251	7.92	-	-	Single Monod	Okabe et al. (1995)
SRBc SRBI	Lactate	0.0036	0.1427	-	-	-	-	-	Dual Monod	Torner-Morales et al. (2010)
SRB	Lactate	*0.00045*	*0.015*	*0.12*	-	*4.9*	-	*0.04*	Single and dual Monod	D'Acunto et al. (2011)
SRBc SRBI	Lactate								Dual Monod	Mattei et al. (2014)
SRB	Lactate	-	0.60	-	-	4.8	-	-	Contois	Oyekola et al. (2012)
Desulfovibrio alaskesnis 6SR	Lactate	-	18.5	0.57	KH$_2$S=0.680 ±0.1 KCd=0.0019	3.12	-	0.192	Levenspiel inhibition model	López-Pérez et al. (2013)
Mixed microbial biomass	Lactate	0.93	-	-	-	-	2.352[a]	-	Single Monod	Xu et al. (2014)
Cv		1.405	1.902	0.969		0.371				

a – expressed as g SO$_4^{2-}$.g^{-1} VS.d^{-1}; Cv – coefficient of variation; Minimum and maximum in italic.

TABLE 2.7 Kinetic parameters for SRB with hydrogen as substrate.

Microbial groups	Substrate (I)	Ks (g SO_4^{2-}.L^{-1})	Kj (g.L^{-1})	Y (g biomass. g^{-1} substrate)	Ki (g.L^{-1})	μ (d^{-1})	ν (d^{-1})	Kd (d^{-1})	Kinetics	References
SRB	Hydrogen	0.0009	0.00005	0.077	K$_{H_2S}$=0.55	5	-	0.03	Dual Monod	Kalyuznyi and Fedorovich (1998)
SRB	Hydrogen	0.0192	0.00007	0.05	K$_{H_2S}$=0.55	2.8	-	0.06	Single and dual Monod	Kalyuznyi et al. (1998)
SRB	Hydrogen	0.018	0.000032	0.0610	-	1.85	-	-	Dual Monod	Spanjers et al. (2002)
SRB	Hydrogen	0.0099	0.0001	0.0366	K$_{H_2S}$=0.2652	-	0.9772	0.01	Dual Monod	Fedorovich et al. (2003)
SRB	Hydrogen	-	0.02625	-	-	-	-	-	Modified Monod	Ribes et al. (2004)
ASRB	H2 and Co2	0.00045	0.00006	0.09	-	1.1		0.04	Dual Monod	Esposito et al. (2009)
SRB	Hydrogen	0.001[a]	0.000022	0.05	-	0.3		0.01	Dual Monod	Tang et al. (2012)
Cv (Hydrogen)		0.828	2.414	0.295		0.73359		0.6324		

a – expressed as gCODL^{-1}; b – expressed as g.L^{-1}.d^{-1}; c – expressed as gProtein.gEthanol^{-1}; d – expressed as g.g^{-1}VSS.s^{-1}; e – expressed as g SO_4^{2-}.g^{-1}VSS.s^{-1};
Cv – Coefficient of variation; Minimum and maximum in italic.

TABLE 2.8 Kinetic parameters for SRB with butyrate, propionate, ethanol and formate as substrates.

Microbial groups	Substrate (i)	Ks (g SO_4^{2-}.L⁻¹)	Ki (g.L⁻¹)	Y (g biomass. g⁻¹ substrate)	Ki (g.L⁻¹)	μ (d⁻¹)	ν (d⁻¹)	Kd (d⁻¹)	Kinetics	References
SRB	Butyrate	0.01	0.009	0.03	K_{iH_2S}=0.4	0.22	-	0.035	Single and dual Monod	Kalyuzhnyi et al. (1998) Poinapen and Ekama (2010)
SRB	Butyrate	0.02016	0.10	0.0329	K_{iH_2S}=0.27642	-	0.45073	0.01	Dual Monod	Fedorovich et al. (2003)
SRB	Propionate	0.0074	0.295	0.035	K_{iH_2S}=0.285	0.81	-	0.018	Dual Monod	Kalyuzhnyi and Fedorovich (1998)
SRB	Propionate	0.019	0.015	0.03	K_{iH_2S}=0.22	0.29	-	0.035	Single and dual Monod	Kalyuzhnyi et al. (1998) Poinapen and Ekama (2010)
SRB	Propionate	0.0192	0.11	0.0329	K_{iH_2S}=0.27642	-	0.41454	0.01	Dual Monod	Fedorovich et al. (2003)
Cv (Propionate)		0.362	0.830	0.063				0.496		
SRB	Ethanol	0.816	0.207	0.0052[b]	K_{EtOH}=80.5 K_{Ac}=7.08	0.312	-	-	Dual Monod	Nagpal et al. (2000)
SRB	Ethanol	-	0.18	-	$K_{Fe^{2+}}$=0.709	-	0.25[c]	-	Single Monod and non-competitive inhibition	Gonzalez-Silva et al. (2009)
SRB[j]	Ethanol	0.091	0.0026	0.076	-	0.019	-	0.01	Dual Monod	Rodriguez et al. (2011)
Cv (Ethanol)		0.629	0.698				-			
SRB	Formate	-	-	0.08 ±0.02	-	-	1.98[a]	-	Monod	Gupta et al. (2011)
SRB	Sulphate	1.34x10⁻⁴	-	-	-	-	3.73x10⁻⁵[d]	-	Monod with molecular diffusion	Overmeire et al. (1994)

a – expressed as g.L⁻¹.d⁻¹; b – expressed as gProtein.g SO_4^{2-}; c – expressed as g_j.g⁻¹VSS.d⁻¹; d – expressed as g SO_4^{2-}.g⁻¹VSS.s⁻¹;
Cv – Coefficient of variation; (Minimum and maximum in italic).

2.2.6 EVALUATION ON MODELS FOR BIOLOGICAL SULPHATE REDUCTION PROCESSES

Several models for biological sulphate reduction have been developed in the last few years with different objectives, such as, understanding biofilm dynamics, microbial competition and inhibition as well as design of bioreactors (Tables 2.1-2.4). In particular, these models are of great importance to understand and test hypotheses of the processes taking place in biofilms on microscale, the population dynamics, to optimize operational performance and to design bioreactor controllers. Despite the great complexity of anaerobic sulphate reducing processes, the feasibility to describe their essential characteristics and dynamics seems evident as already done in the models of Table 2.4. Processes such as substrate degradation and accumulation, microbial interaction and growth can have great impact on process control systems (see section 2.4.2). Thus, more knowledge and information is required to fully understand such processes in order to accomplish more complete and accurate models. See for instance Klok et al. (2012) and Klok et al. (2013) for the introduction of physiologically based kinetic models for bacterial sulphide oxidation. The models developed are very specific in their nature and thus, not simple to be adapted to a full scale application. Thus, it is advisable to develop a more generic model, such as the ADM, comprising the different processes encountered in previous sections for sulphate reduction bioreactors so that the gained knowledge can be easily transferred to others.

The choice of model is directly dependent on the defined goals and underlying processes. If several substrates are used and thus, microbial competition is expected one might consider using a similar approach to Fedorovich et al. (2003) or Poinapen and Ekama (2010). When several microbial groups and limited substrates are present, it is advisable to use models similar to Esposito et al. (2009) or Frunzo et al. (2012). Inhibition can sometimes play a big role in such systems and if so, metal inhibition (Gonzalez-Silva et al., 2009), acetate and/or sulphide (Reis et al., 1992; Moosa and Harrison, 2006) should be included in the model.

On the other hand, if the model is to be used in a control system (see section 2.4), for more or less time-invariant processes, i.e., dynamic systems with constant rate coefficients, then it is advisable to reduce its complexity to a minimum so that it is able to simulate and predict the response of the bioreactor to different events on a short time scale. It is important to note, that in general, the more variables to control the more complex the model should be as a result of the interactions between the variables. In such cases, it is common to make gross simplifying assumptions, which may be eliminated or improved as knowledge increases. Critical judgement must be used in order to minimize the errors associated to these simplifications. Therefore, the theoretical assumptions, choice of model parameters and accuracy of the numerical solution method are crucial to obtain valid models (Dunn et al., 2003). Moreover, sensitivity analysis and cross-validation techniques, as in Keesman (2011), will help to find invalid assumptions and incorrect descriptions of subprocesses. In addition, one might consider using ANN or other types of black-box models (see section 2.4.3) which require less prior information on structure and interaction between variables when compared to mechanistic models.

2.3 PROCESS MONITORING

This review describes chemical sensors, microsensors and biosensors used or that may be used for the monitoring of sulphate reduction processes. In the first part, sensors for chemical analysis of the microbial activity are reviewed. The second part focuses on how the sensor measurements can be combined with molecular techniques to determine both the activity and microbial ecology in bioreactors where sulphate reduction occurs. Many traditional laboratory-based analytical techniques are commonly used to measure crucial parameters for monitoring sulphate reduction processes due to their great reproducibility and precision. However, these techniques are mostly off-line, time consuming and require extensive manual handling. In the past decades, much effort and research has been put in the development of real-time monitoring equipment. An overview of these real-time monitoring techniques to measure sulphide, electron donors, sulphate and biomass composition is given below.

2.3.1 *IN SITU* SULPHIDE SENSORS

Dissolved sulphide measurements are very important in sulphate reduction processes, because sulphide is the end-product of the process. Hence, dissolved sulphide measurements are frequently used to measure the efficiency of the process. The classical off-line methods for sulphide measurements, e.g. methylene blue, Cord-Ruwisch or other iodometric methods (Cord Ruwisch, 1985; APHA, 1995) are rather time-consuming and require elaborate sample handling. To overcome this, Ion Selective Electrodes (ISE) or similar electrode types (such as the Ag/Ag_2S) have been developed. ISE are usually chosen for routine applications due to the fact that they have many advantages over other methods for ion concentration determination. These include analysis speed, portability, no sample destruction and wide measuring range. In ISE, selectivity is introduced by the receptor molecules (or ionophores) which are usually immobilised in a polymeric membrane matrix. The receptor molecule attributes selectivity to the sensor by its strong and selective interactions with the target analyte (Morigi et al., 2001). From the measured activity of the free sulphide ions with an ISE, the analytical concentration of the total dissolved sulphide (TDS) can also be calculated if the protonation constants of the sulphide ion (K_1 and K_2) and the pH of the sample solution are known:

$$S^{2-} = \frac{TDS}{1 + \dfrac{(H^+)}{K_2} + \dfrac{(H^+)^2}{K_2 K_1}} \qquad (2.23)$$

Grootscholten et al. (2008) used equation (2.23) to estimate simultaneously the sulphide and metal concentrations in a precipitation reactor using an on-line estimation algorithm, also called a software sensor. The zinc concentration and precipitation rate in the CSTR were estimated based on the pH and pS (which measures the activity of the S^{2-} species and is defined as: $-log[S^{2-}]$) in the reactor.

Table 2.9 summarizes the characteristics of sensors that measure sulphide concentrations, such as concentration range, pH range and tested interfering compounds. Frevert and Galster (1978) suggested a combined pH glass and sulphide electrode measuring system for the direct determination of the total sulphide concentration in solution. However, this system was only developed for pH < 5, while the pH of the sulphide containing natural waters and sewage is usually higher, i.e., pH > 7. Guterman et al. (1983) and Tóth and Solymosi (1988) developed an appropriate microprocessor interface for the pH ranges of 7.5 - 11.5 and 9 - 12, respectively, using a sulphide ISE as sensor. In the work of Guterman et al. (1983), the authors were able to measure sulphide in the concentration range of 10^{-5} to 10^{-1} M. To determine the total sulphide concentration for sewage waters in the pH range 3 - 11.4, an electrochemical method was tested using a potentiometric cell, which contained either a sulphide ion selective ISE-glass electrode pair or an Ag/Ag_2S electrode-glass electrode system (Schmidt et al., 1994). Both gave good results for the measurement of the total dissolved sulphide, in a sulphide concentration range of 10^{-12} to 10^{-2} M. Villa-Gomez et al. (2014) showed that an Ag_2S pS electrode could be used to continuously monitor on line the sulphide concentration in sulphate reducing bioreactors and can thus be used to develop a strategy for sulphide control (See section 2.4.2).

Recently, an electrode with a modified PVC membrane with surfactant modified clinoptilolite zeolite was applied successfully for the measurement of sulphide in wastewater samples (Nezamzadeh-Ejhieh and Afshari 2012). It showed a good response for sulphide concentrations between 10^{-7} and 10^{-1} M with a detection limit of $6.6x10^{-8}$ M. It also showed good performance for the pH range 3 - 10. Other types of sensors are the chalcogenic glass chemical sensors for S^{2-} and dissolved H_2S. These are good for the detection in a broad pH range (5 - 11) and exhibit better sensitivity, enhanced selectivity (no notable effect in the presence of Cl^-, NO_3^- and SO_4^{2-}) as well as response stability at neutral pH compared to commercial sulphide ion sensors (Miloshova et al., 2003). The sulphide concentration range is dependent on the glass membrane composition. However, the authors did not disclose the details due to patent related issues.

As an alternative method to measure dissolved sulphide concentrations in sulphide oxidizing bioreactor systems, a redox electrode was proposed (Janssen et al., 1998). The redox potential is mainly determined by the sulphide concentration since it has a high standard exchange current density with the platinum electrode surface. Thus, by maintaining a particular redox set-point value, the reactor becomes "sulphide-stat". In contrast with the previously discussed sensors, the redox potential reading depends less on pH fluctuations of the solution.

At neutral pH a great part of the sulphide will be present as H_2S, which is easily transferred into the gaseous form. The measurement of the latter can be done by numerous sensors. This has been extensively reviewed in the work of Pandey et al. (2012). One example of a H_2S sensor is the wireless electronic nose system (WENS). Electronic noses usually consist of an array of sensors for chemical detection, a data acquisition system and a mechanism for pattern recognition, such as neural networks or neuro-fuzzy networks. WENS showed good performance in the concentration range of 0.15 – 1.5 ppm H_2S (Cho et al., 2008). The sensor elements and the electronics are integrated in a chip, thus increasing the sensibility and decreasing the measurement time (0.2 s), which is very important for automatic control. Thus, WENS seem to be an attractive option for real-time monitoring of H_2S at the sub-ppm concentration range. However, these sensor arrays and/or electronic noses are still at the development stage, and more research on their application in process control is ongoing.

The quantification of sulphate reduction rates within biofilms was only possible after performing mass balance calculations or by tracer techniques in which biofilms were growing on metal surfaces that react with the produced H_2S (Kühl and Jørgensen, 1992). The development of microsensors, which can monitor the processes within biofilms, was a great improvement in this field. Kühl and Jørgensen (1992) successfully used oxygen, pH and sulphide microelectrodes to study the microzonation and dynamics of oxygen respiration, sulphide oxidation and sulphate reduction at high spatial resolution in aerobic biofilms collected from a trickling filter. The calibration curves for the sulphide microelectrodes exhibited a log linear response for 10^{-3} to 10^{-6} M H_2S. The electrode response times depend on the H_2S concentration and varied from < 1 min for the highest concentration up to 10 - 15 min for the lowest.

The construction and use of well-functioning Ag/Ag_2S electrodes can, however, be problematic (Kühl et al., 1998) due to e.g. non-ideal responses, signal drift and very long response times at low sulphide levels. The very high pK_2 of the sulphide system may also prevent the application of such electrodes in acidic environments, where S^{2-} is practically inexistent. Kühl et al. (1998) optimized a H_2S microsensor based on the amperometric measuring principle. The microsensor allowed to obtain a microprofile of H_2S in an acidic lake sediment with a flocculant surface layer several cm thick (Figure 2.8). Its application was demonstrated for sulphate reduction and sulphide oxidation studies in acidic sediments. The microsensor exhibited a fast (0.2 to 0.5 s) and linear response over a concentration range of 1 to > 1000 µM H_2S at pH 4.6 and relatively low SO_4^{2-} concentrations. This microsensor also showed good results for neutral and moderate alkaline (pH < 9) biofilms and sediments. The microelectrode was used to measure the H_2S concentration and to quantify the microbial sulphate reduction activity in a process that removes uranium (U(VI)) (Beyenal et al., 2004).

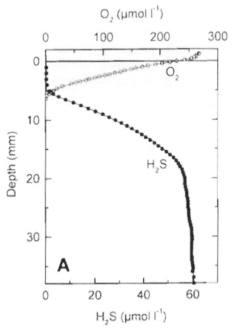

FIGURE 2.8 Oxygen and H_2S concentration profiles measured in acidic (pH 4.6) sediment from Lake Fuchskuhle, Germany (Kühl et al,. 1998).

TABLE 2.9 Selected electrodes for sulphide measurements.

Measuring principle	Concentration range	Application	pH	Interfering anions	Non-Interfering anions	References
Combined pH glass and ISE	-	-	<0.5	-	-	Frevert and Galster (1978)
Microprocessor interface for ISE	10^{-5} to 10^{-1} M	-	7.5-11.5	-	-	Guterman et al. (1983)
Microprocessor interface for ISE	10^{-5} to 10^{-1} M	-	9.0-12-0	-	-	Tóth et al. (1988)
ISE-glass electrode pair or an Ag/Ag_2S electrode-glass electrode system	10^{-12} to 10^{-2} M	Total sulphide concentration in sewage waters	3.0-11-4	-	-	Schmidt et al. (1994)
PVC membrane with surfactant modified clinoptilolite zeolite	10^{-7} to 10^{-1} M	potentiometric titration of S^{2-} with Zn^{2+}, Cu^{2+}; direct determination of S^{2-} in waste water samples	3-0-10.0	NO_3^-, ClO_3^-, Cl^-, Br^-, CH_3COO^-, SP_3^{2-}, I^-, CN^-	$C_2O_4^{2-}$, PO_4^{3-}, AsO_3^{3-}, ClO_4^-	Nezamzadeh-Ejhieh and Afshari (2012)
Chalcogenic glass chemical sensors	-	-	5-0-11.0	CO_3^{2-}	Cl^-, NO_3^-, SO_4^{2-}	Miloshova et al. (2003)
Redox electrode	-	Sulphide oxidation control	8.0	-	-	Janssen et al. (1998)
Microelectrode	10^{-3} to 10^{-6} M H_2S	Understanding microzonation and dynamics of sulphide oxidation and sulphate reduction in aerobic biofilms from trickling filters	-	-	-	Kühl and Jørgensen (1992)
Microelectrode based on the amperometric measuring principle	1 to >1000 μM H_2S	Understanding sulphate reduction and sulphide oxidation in acidic sediments	<9.0	-	-	Kühl et al. (1998)

-information not available

2.3.2 *IN SITU* SENSORS FOR ELECTRON DONOR CONCENTRATIONS

The real-time measurement of electron donors the SRB use is very helpful when automated control of the process is wanted. Indirect measurements of most of the electron donors (lactate, volatile fatty acids) use the chemical oxygen demand (COD). In recent years, the development of rapid and environmentally friendly methods for the COD determination has attracted more attention. Most of these new methods were based on electrocatalysis using PbO_2 (Ai et al, 2004a; Li et al., 2005), Cu/CuO (Silva et al., 2008), boron-doped diamonds by amperometric methods (Yu et al., 2007) or ultrasound electrodes (Wang et al., 2012), photocatalysis (PC) and photoelectrocatalysis (PEC) with TiO_2-based materials (Kim et al., 2000; Ai et al., 2004b; Zhao et al., 2004; Chen et al., 2005; Li et al., 2006; Zhu et al., 2006; Zheng et al., 2008; Zhang et al., 2009a; Zhang et al., 2009b; Qu et al., 2010). Each electrode type shows some limitations, mostly due to the presence of interfering compounds. Hence, special attention must be given to these when choosing the appropriate electrode.

Hydrogen can also be used as an electron donor in sulphate reduction processes. There are many different types of hydrogen sensors commercially available or under development (Hübert et al., 2011). These sensors are based on different measurement principles such as catalytic, thermal conductivity, electrochemical, mechanical, optical, acoustic resistance or work function. Most of these sensors were developed to detect and measure hydrogen in the gas phase. Thus, the determination of dissolved hydrogen is only possible through calculations, which are the basis of the so called software sensors. A Si-based combined chemosensor capable of simultaneous amperometric/field-effect detection has been used to quantify dissolved hydrogen directly for biogas applications (Huck et al., 2012).

2.3.3 *IN SITU* SULPHATE SENSORS

Many sensors for the detection of sulphate have been developed and tested in laboratory conditions. Each is applicable to a specific sulphate concentration and pH range and thus the user must take this into account when selecting the appropriate sensor. Although some of the later studies present sensors applicable in a wide sulphate concentration and pH range, to our knowledge, there is no commercial on-line sensor available for accurate detection of sulphate in wastewaters. In addition, there are still few studies on the application of either chemical or biological sensors for sulphate monitoring in wastewaters. The selectivity of a sensor towards a targeted compound, sulphate in this case, is also of great importance. This must be taken into consideration when choosing an appropriate sensor. Table 2.10 overviews the characteristics to take into account as well as tested interfering compounds.

A number of ion-recognition elements have been proposed for the development of a sulphate ISE. The response of an ISE should, in theory, be in accordance with the Nernst equation:

$$E = const - 2.303 \frac{RT}{z_i F} \log(a_i) \qquad (2.24)$$

where E is the electrode potential, R is the universal gas constant, T is the temperature, F is the Faraday constant, z_i is the ionic charge and a_i is the activity of the ion.

A bisthiourea ionophore ISE showed a Nernstian response, at pH 7.0, for concentrations ranging from 10^{-6} to 10^{-2} M and also presented acceptable selectivity for sulphate ions in comparison with SCN^-, NO_3^- and Br^- (Nishizawa et al., 1998). The concentration range for bisthiourea ionophore ISE was enlarged to 3.0×10^{-7} to 1.0×10^{-1} M in the work of Firouzabadi et al. (2013). By adding cetyltrimethylammonium bromide and various plasticizers, good sensitivity with respect to many common anions was achieved. A Schiff base Zn(II) complex also revealed a Nernstian response (Shamsipur et al., 2001), but only in the 5.0×10^{-5} and 1.0×10^{-1} M concentration range. For the latter, the potentiometric response does not depend on the pH of the solution in the pH range 3.0 - 7.0 and assisted in the potentiometric titration of sulphate and barium ions. The response of an ISE with an imidazole derivative (Li et al., 1999) was closely related to the pH of the solution. A linear response was obtained, at pH 3, in the concentration range of 3.2×10^{-5} to 0.5 M and was applied for the determination of sulphate in pharmaceutical samples. A zwitterionic bis(guanidium) ionophore bearing a dihydrochloride analogue ISE was investigated and showed a Nernstian behaviour in a concentration range of 10^{-6} to 10^{-2} M in the presence of Cl^- concentrations below 10^{-3} M (Fibbioli et al., 2000). On the other hand, a tris(2-aminoethylamine) derivative ISE (Berrocal et al., 2000) showed a Nernstian response for higher sulphate concentrations (10^{-5} to 10^{-1} M).

An ISE based on the dispersion of hydrotalcites into a poly(dimethylsuloxane) membrane had linear response in the sulphate range of 4.0×10^{-5} to 4.0×10^{-2} M, which was constant over a pH range 4.0 - 7.0 (Morigi et al., 2001). The ISE was successfully applied in the sulphate determination in commercial mineral waters. A PVC-membrane ISE based on 2,5-diphenyl-1,2,4,5-tetraaza-bicyclo[2.2.1]heptane as a neutral carrier revealed a linear response for sulphate concentrations ranging from 9.0×10^{-6} to 1.0×10^{-1} at a pH of 4.0 (Shamsipur et al., 2002). A derivative of pyrilium perchlorate was also used as a neutral carrier for a PVC-membrane ISE. The range of detection was slightly less, 1.0×10^{-6} to 1.0×10^{-2} M, but it was less sensitive to pH changes, working in a pH range 4.0 - 9.0 (Ganjali et al., 2002). This sensor was applied as an indicator electrode in the potentiometric titration of sulphate and barium ions in aqueous solutions with varying anions (Table 2.10) and for the indirect determination of the zinc concentration in zinc sulphate tablets.

An electrode with a surfactant modified zeolite carbon paste was applied for the potentiometric determination of sulphate (Nezamzadeh-Ejhieh and Esmaeilian, 2012). It showed a good nernstian response for concentrations between 2.0×10^{-6} and 3.1×10^{-3} M and with constant nernstian response for pH 4 - 10. The electrode was applied to determine sulphate concentrations in a pharmaceutical zinc sulphate capsule (78.53 ± 1.53 mg SO_4^{2-}. mg capsule^{-1}) and as an indicator electrode in the potentiometric titration of sulphate. Other neutral carriers used are zinc-phthalocyanine (Ganjali et al., 2003), 1,3,5-triphenylpyrylium perchlorate (Ganjali et al., 2004) and 2-amino-6-(tbutyl)-4-(pyridine-2-yl)pyrimidine) (dichloride)palladium(II) (Mizani and Rajabi, 2014). These three carriers presented good linear responses for sulphate concentrations between 1.0×10^{-6} and 1.0×10^{-2} M (pH 2.0 - 7.0), 6.3×10^{-6} to 1.0×10^{-1} (pH 2.5 - 9.5) and 5.0×10^{-1} to 4.0×10^{-7} (pH 2.9 - 9.5), respectively.

A biosensor for sulphate was developed using *Thiobacillus ferrooxidans* strain 15 for the measurement of SO_4^{2-} in acid rain (pH range 2-3) (Sasaki et al., 1997). In this microbial sensor, the bacteria used oxidize Fe(II) in the presence of sulphate. The sulphate concentration is calculated based on the decrease of current at the microbial electrode induced by the oxidation of Fe(II) to Fe(III) and simultaneous consumption of dissolved oxygen. The biosensor showed linear responses between 4 and 200 µM SO_4^{2-}. However, nitrate was an interfering substance for this sensor, which also showed poor stability.

TABLE 2.10 Selected electrodes for sulphate measurements.

Measuring principle	SO_4^{2-} concentration range (M)	pH range	Interfering anions*	Non interfering anions	References
Bisthiourea ionophore ISE	10^{-6} to 10^{-2}	7.0	Br^-, NO_3^-, NO_2^-, SCN^-	Cl^-, CH_3COO^-, SO_3^{2-}	Nishizawa et al. (1998)
Derivative of imidazole ISE	3.0×10^{-7} to 1.0×10^{-1}	3.0-9.0	-	F^-, Br^-, I^-, Cl^-, NO_3^-, SCN^-, ClO_4^-, I_3^-, CN^-, HPO_4^{2-}, CH_3COO^-, Citrate, NO_2^-, $C_2O_4^{2-}$, ClO_4^-	Firouzabadi et al. (2013)
	3.2×10^{-5} to 0.5	3.0	Br^-, NO_3^-, NO_2^-	Cl^-, CH_3COO^-	Li et al. (1999)
Zwitterionic bis(guanidium) ISE	10^{-6} to 10^{-2}	-	Br^-, I^-, NO_3^-, ClO_4^-, CO_3^{2-}	Cl^-, CH_3COO^-	Fibbioli et al. (2000)
Tris(2-aminoethylamine) derivative	10^{-5} to 10^{-1} M	-		Br^-, I^-, Cl^-, NO_3^-, SCN^-, ClO_4^-, SO_4^{2-}, SO_3^{2-}	Berrocal et al. (2000)
Schiff base complex of Zn (II) ISE	5.0×10^{-5} to 1.0×10^{-1}	3.0-7.0	-	Br^-, I^-, Cl^-, NO_3^-, NO_2^-, CN^-, SCN^-, ClO_4^-, CH_3COO^-, SO_3^{2-}, CO_3^{2-}	Shamsipur et al. (2001)
Dispersion of hydrotalcites ISE	4.0×10^{-5} to 4.0×10^{-2}	4.0-7.0	NO_3^-, SCN^-	Br^-, Cl^-, CH_3COO^-, $H_2PO_4^-$	Morigi et al. (2001)
2,5-diphenyl-1,2,4,5-tetraa-za-bicyclo[2.2.1]heptane ISE	9.0×10^{-6} to 1.0×10^{-1}	4.0	-	Br^-, I^-, Cl^-, NO_3^-, NO_2^-, CN^-, SCN^-, ClO_4^-, CH_3COO^-, SO_3^{2-}, CO_3^{2-}	Shamsipur et al. (2002)
Derivative of pyrilium perchlorate ISE	1.0×10^{-6} to 1.0×10^{-2}	4.0-9.0	SCN^-	Br^-, I^-, Cl^-, NO_3^-, NO_2^-, CN^-, ClO_4^-, SO_3^{2-}, CO_3^{2-}	Ganjali et al. (2002)
Zinc-phthalocyanine ISE	1.0×10^{-6} to 1.0×10^{-2}	2.0-7.0	-	Br^-, Cl^-, NO_3^-, NO_2^-, SCN^-, ClO_4^-, CH_3COO^-, SO_3^{2-}, CO_3^{2-}, $H_2PO_4^-$	Ganjali et al. (2003)
1,3,5-triphenylpyrylium perchlorate ISE	6.3×10^{-6} to 1.0×10^{-1}	2.5-9.5		I^-, Cl^-, NO_3^-, NO_2^-, CN^-, SCN^-, ClO_4^-, CH_3COO^-, SO_3^{2-}, $H_2PO_4^-$	Ganjali et al. (2004)
2-amino-6-(tbutyl)-4-(pyridine-2-yl)pyrimidine)(dichloride) palladium(II)	5.0×10^{-1} to 4.0×10^{-7}	2.9-9.5		I^-, Cl^-, NO_3^-, NO_2^-, SCN^-, ClO_4^-, CH_3COO^-, SO_3^{2-}, HPO_4^{2-}, Br^-, F^-, Tartrate^{2-}	Wizani and Rajabi (2014)
Microbial sensor using *Thiobacillus ferrooxidans*	4×10^{-6} to 2.0×10^{-4}	2.0-3.0	Cl^-(>8.97mM); NO_3^- (>200µM)	-	Sasaki et al. (1997)

*Anions were considered to be interfering when the selectivity coefficient was equal or higher to 1; - : information not available

2.3.4 SENSORS FOR BIOMASS COMPOSITION

The microsensors described in the section 2.3.1 can be coupled to molecular techniques to get more insight in the processes prevailing in a biofilm. This combination was done for the first time by Ramsing et al. (1993) to study SRB in trickling-filter biofilm treating municipal wastewater. Gradients of O_2, H_2S and pH were measured with microelectrodes and the distribution of SRB was determined by specific oligonucleotide probes. This approach of using microelectrodes together with specific oligonucleotide probes proved fruitful in that it was possible to relate the distribution of bacteria to their chemical microenvironment at a spatial resolution of below 100 μm.

Analysing the transients of sulphate reduction, using microsensors, and the successive changes in the composition of microbial species using molecular techniques (PCR, DGGE) were studied in a multi-species aerobic bacterial biofilm (Santegoeds et al., 1998). The goal of this study was to determine how the species composition is related to the activity in a biofilm with microenvironments changing gradually. However, the molecular techniques used were not sufficient to accurately predict the microbial population changes in this complex environment. Other molecular techniques (DGGE, PCR and FISH) in combination with microsensors for H_2S and CH_4 were used to study the activity distribution in anaerobic aggregates and the population structure (Santegoeds et al., 1999). The microsensors and molecular techniques used provided direct information about sulphate reduction and methanogenesis in UASB aggregates (Figure 2.9). The data obtained on the community structure could then be related to the metabolic function of the respective populations.

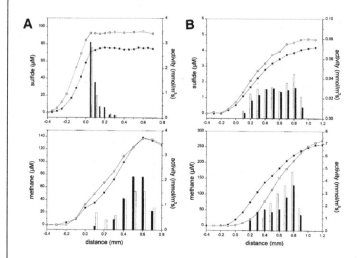

FIGURE 2.9 Sulphide and methane microsensor profiles (lines) and activity values (bars) in methanogenic-sulphidogenic (A) and methanogenic (B) aggregates in the presence (white) or absence (black) of sulphate. The aggregate surface is at distance of 0 mm, the centre of the aggregates is at a distance of ca. 0.9 mm (Santegoeds et al,. 1999).

Microsensors with a high spatial resolution were used to measure O_2 and H_2S profiles and to localise aerobic respiration and sulphate reduction activities within the biofilm (Kühl et al., 1998). The molecular techniques used included DGGE and PCR-amplified 16S ribosomal DNA fragments to determine the microbial complexity in the biofilm in an acidic lake sediment. The researchers were able to follow the development of the microbial community and to detect several SRB groups in complex biofilms with several species. However, the techniques also showed some limitations such as the inability to quantify the activity and the difficulty in identifying the specific molecular probes for the species present.

The techniques for biomass and activity characterization mentioned above have the disadvantage of being invasive, destructive and do not give information in real time. Some progress has been made to quantify microbial activity with online monitoring. An increasing trend towards the development of impedimetric biosensors is observed. Impedimetric biosensors have been fabricated to study biomolecular reactions (Oliveira et al., 2008) as well as specific recognitions of proteins (Bogomolova et al., 2009), lectins (La Belle et al., 2007), antibodies (Rezaei et al., 2009) or nucleic acids (Hu et al., 2011). For SRB detection, rapid and non-labelled impedimetric biosensors were developed based on agglutination reactions (Wan et al., 2009), antibody recognition platforms on 3D Ni foam substrates (Wan et al., 2010a), self-polymerized polydopamine films (Wan et al., 2011b), RGSs-CS nanocomposite films (Wan et al., 2011a) and on bioimprinted films (Qi et al., 2013a). An electrochemical SRB detection method based on the conversion of ZnO to ZnS nanorods arrays by the sulphuration process was developed and showed promising results (Qi et al., 2013b). An *in situ* methodology based on covalently bound redox indicators can be used for determining when sulphate-reducing conditions exist in environmental samples. Sulphide coupled well to the cresyl violet immobilized redox indicator in the concentration range of 1 – 100 mM total sulphide and the pH range of 6 – 8. Thionine, the indicator with the highest potential (actual potential measured by the electrode), reacts rapidly with sulphide at levels well below 1 mM (Jones and Ingle Jr, 2005). The amplification of responses of vancomycin-functionalised magnetic nanoparticles, using a quartz crystal microbalance under an external magnetic field, gave good results to detect and quantify SRB (Wan et al, 2010b). Potentiometric stripping analysis was used for the selective detection of *Desulfovibrio caledoiensis* (Wan et al., 2010c).

Another easily measurable parameter to quantify the microbial activity is conductivity. The metabolism of bacteria can cause an increase in the conductivity of the culture medium due to the generation of charged, mobile metabolites such as organic acids and the decomposition of large molecules into smaller ones. The changes in the conductivity can then be correlated with bacterial activity and be used to enumerate bacteria. Lyew and Sheppard (2001) used conductivity measurements to measure the SRB activity for the treatment of acid mine drainage. They concluded that the latter is more sensitive for the assessment of SRB activity than pH or the oxidation-reduction potential.

2.3.5 EVALUATION ON PROCESS MONITORING

On line monitoring of substrates, products and possible intermediates leads to increased knowledge on the process and thus, more accurate models and controller applications are possible. Although there has been great advance in the development of sensors, there are still few reports on their application on continuous sulphate reducing bioreactors. A successful example was the use of a solid state Ag_2S ion selective electrode assisting in a control strategy design for biological sulphide production in bioreactors (Villa-Gomez et al., 2014). Several online sensors have been developed for the measurement of crucial variables in the sulphate reduction processes. The development of these sensors is bringing researchers one step closer to a better understanding of the process and to validate models developed, for example, for the dynamics inside biofilms (Mašić et al., 2010). These type of sensors seem to be a promising alternative to the traditional methods of monitoring chemical substances and microbial populations. However, more research is needed for the optimization of the discussed techniques in order to minimize the interference with other wastewater contaminants and minimize response times so that they can be utilized in control strategies for continuous bioreactors treating wastewater.

Some sensors are already commercially available, but many of them are still under-developed. Especially for sulphate, for which there is still no published research on (micro) sensors tested in wastewater. The latter would be very interesting to use in control of the sulphate load to a bioreactor or to experimentally study the substrate diffusion in biofilms.

Overall, no sensor has presented optimal overall performance and thus, the choice must be directly related to the characteristics of the process to be monitored, i.e., concentration range, pH, interfering compounds (Tables 2.9-2.10), cost, ease of use, placement of the sensors, response time, reliability, accuracy and detection limit (Bourgeois et al., 2001). However, as shown in this review, promising research is being done in the development and optimization of these on-line measuring devices and thus, allowing further optimization in the monitoring step of bioprocess controllers.

2.4 CONTROL OF ANAEROBIC SULPHATE REDUCTION PROCESSES

The design of advanced bioprocess control strategies is highly related to the available models and sensors. With the development of models and *in situ* sensors for online monitoring, it is now possible to develop high performance control strategies to control biological systems such as sulphate reducing bioreactors. This section reviews the existing (Table 2.11) and potential control strategies for sulphate reducing bioreactor systems.

Large progress has been made in recent years in the control of anaerobic digestion processes. In these processes, the controlled variables are usually the process intermediates such as the volatile fatty acid concentration, pH, bicarbonate alkalinity or gas concentration flow rates (Pind et al., 2003). However, the number of experimental applications of control approaches in sulphate reducing bioreactors is still scarce. Mathematical models, presented in the section Models for biological sulphate reduction processes, can serve as support for the design of control strategies. When selecting a control strategy one must take into account the unique characteristics of the process to be controlled. In the studies presented in this section, single input-single output feedback control is used. Feedback control (Figure 2.10) starts by measuring the variable to be controlled and comparing it to the set-point on a set-point trajectory, defined by the user. It then uses the difference between these two values to determine which action to be taken by the controller that will then change the manipulated variable (Dunn et al., 2003).

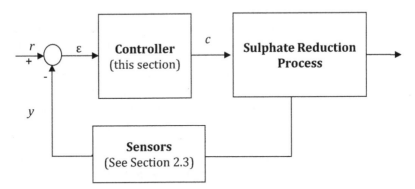

FIGURE 2.10 Feedback control loop for simple process control of sulphate reduction. Where r is the reference input, ε is the tracking error, c is the manipulated variable and y is the sensor output variable.

The sections below review control strategies utilized for different sulphate reduction and anaerobic digestion processes. The first section will focus on the control of chemical/biological sulphide addition, the second section on the control of biological sulphide production and the third will evaluate the use of adaptive controllers for sulphate reduction bioprocesses.

2.4.1 CONTROL OF SULPHIDE ADDITION

Simple control strategies can be chosen when the process to be controlled presents itself with low complexity. In these cases, a commonly used controller is the so-called Proportional-Integral (PI) controller. A PI controller has two adjustable parameters, the controller gain, K_c, and the integral time, τ_i. These parameters can be obtained by using different tuning methodologies. More information on these tuning methodologies can be found in the literature (Dunn et al., 2003).

Selective metal precipitation with sulphide has been shown possible by applying a combination of a pS and pH electrode, and controlling the addition of chemical sulphide using a PI control strategy to achieve the stoichiometric addition of sulphide entering a precipitation reactor (Veeken et al., 2003; Esposito et al., 2006; Sampaio et al., 2009; Sampaio et al., 2010). In the work of Veeken et al. (2003), experiments were performed in batch and continuous systems with synthetic wastewater containing Cd, Cu, Ni, Pb and Zn. The heavy metals were successfully removed to concentrations < 0.05 mg.L^{-1} at pH 6.0 by sulphide precipitation, while maintaining the total sulphide concentration < 0.02 mg.L^{-1}. During precipitation, the pS-electrode gave a unique response for each heavy metal. The latter was directly related to the solubility product of the corresponding metal sulphide.

Thus, the metals in mixtures of Cu-Zn (CSTR) and Pb-Zn (batch reactor) were selectively precipitated from solution at pH 6.0 by control of the pS at different levels. At pH 6.0, the pS values for Cu, Pb and Zn were 39.0, 30.0 and 24.0, respectively. This resulted in the production of pure metal sulphide sludges that could possibly be reused.

In a similar work, Sampaio et al. (2009) measured the process variable in the reactor and manipulated the sulphide flow using a feedback control (Figure 2.10). The reactor was run at constant metal and sulphide flows and, at a sudden point, the sulphide flow was increased to another constant value. The pS electrode response to this step change can then be used to calculate the PI controller parameters and . Consequently, it was possible to continuously and selectively precipitate Cu with chemical sulphide to concentrations below 0.3 ppb from water containing around 600 ppm of both Cu and Zn in a CSTR at pH 3 and pS 25. The Cu was recovered with a purity of around 100%, whereas the total soluble sulphide concentration was below 0.02 ppb even with increasing input concentrations (Figure 2.11). Later, Sampaio et al. (2010), using a similar strategy, showed the selective removal of Zn in a CSTR at pH 5 and pS 18 from an aqueous mixture of Zn/Ni with a purity of 99%. The current design of the pS electrode appeared to be incompatible with the NiS precipitation process at pH 4 - 6 due to interferences of the precipitates with the pS electrode.

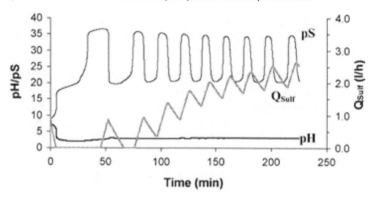

FIGURE 2.11 Continuous selective precipitation of Cu from Zn controlled at pS 25 and pH 3 (Sampaio et al., 2009); (Q_{suf} is the sulphide flow).

The PI control strategy was adapted and successfully used to control biogenic sulphide entering a precipitator reactor with only metal precipitation taking place (König et al., 2006). Similarly, the Cohen Coon method (Stephanopoulos, 1984) was used to estimate the control parameters based on experimental step response data, i.e., based on the sulphide potential response to a step-wise variation of the buffer flow. The pS was controlled at 15 and the pH at 5.87 ± 0.55. The pS/pH control system was able, using a PI controller, to bring the sulphide concentration to the desired value within acceptable margins regarding the optimal removal of zinc. Esposito et al. (2006) assessed the performance of a zinc sulphide precipitation process using a PI control algorithm to control the pH and sulphide (both chemical and biogenic) concentration using pH and pS electrodes. A residual zinc concentration of 0.07 mg/L was obtained from the precipitation of zinc sulphide at pS 15 from a 3 g/L Zn^{2+} influent for both sulphide sources. However, at pS 10 and 20 the ZnS precipitation efficiency decreased when using the biogenic instead of the chemical sulphide, which was related to the presence of other substances present in the sulphate-reducing bioreactor where the sulphide was produced.

2.4.2 CONTROL OF BIOLOGICAL SULPHIDE PRODUCTION

For metal removal and recovery processes, the required amount of sulphide to be produced by SRB depends on the composition of the wastewater to be treated, i.e. its metal concentration. Steering the sulphide production towards this required stoichiometric amount in bioreactors is highly relevant to avoid H_2S overproduction, which increases operational costs and may impose a sulphide removal post treatment step (Villa-Gomez et al., 2014). To control the production of sulphide in a bioreactor is more complex since the production itself must be considered in the control strategy. This strategy must take into account the amount of substrate added and thus, an additional control parameter is required to overcome the lag time between substrate dosing, substrate bioconversion and release of the desired product (S^{2-}) (Villa-Gomez et al., 2014).

In practice, proportional-integral (PI) control is sufficient in most systems. However, when fast changes are anticipated in the process, a derivative (D)-action may be added to smooth the controller response by predicting the error in the immediate future (Dunn et al., 2003; Dunn et al., 2005). In addition, if the D-action is used, typically a first-order filter is added to make the controller strictly proper, i.e., to increase the stability of the controller. The proportional-integral-derivative (PID) controller has been successfully used in anaerobic bioreactors, partly due to the derivative control parameter used to overcome the lag caused by the time required for substrate degradation prior to being used or transformed to the desired product (Dunn et al., 2003; Dunn et al., 2005; Jagadeesh and Sudhaker, 2010). In addition to the controller gain and the integral time, the PID controller has an extra adjustable parameter called the derivative time . More detailed information on PID controller can be found in the literature (Dunn et al., 2003; Dunn et al., 2005).

Different tuning strategies, using a pS electrode, were evaluated to control the production of biological sulphide in an inverse fluidized bed reactor performing biological sulphate reduction. The bioreactor was run on automated operation using the LabView software version 2009® (Villa-Gomez et al., 2014). Step changes in the organic loading rate (OLR) were applied either by changing the HRT or the concentration of lactate in the influent (Figure 2.12). The pS output values resulting from both control strategies were used to determine the PID parameters. As a result, the sulphide concentration in these bioreactors is likely to be controlled by the variation of the lactate concentration or of the HRT depending on the desired outcome of the sulphide concentration. A variation of the lactate concentration should be applied if an increase in the sulphide concentration is desired, and a change in the HRT should be applied if the sulphide concentration is to be decreased. This was a crucial first step in showing that a control strategy for sulphide production towards a desired concentration is possible, but more research is needed to turn this into practical applications. The critical factors when controlling biological sulphide production are the delays in response time, the time variant response and a high control gain (Villa-Gomez et al., 2014). Thus, it is crucial to fully understand the metabolic pathways in the sulphate reducing biomass in order to overcome these setbacks and optimize the design of a control strategy.

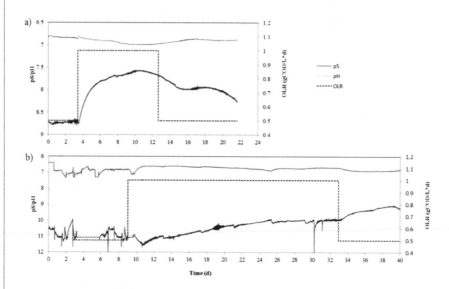

FIGURE 2.12 Step responses of the pS electrode response for a change in a) COD$_{in}$ and b) HRT (Villa-Gomez et al., 2014).

2.4.3 ADAPTIVE CONTROL OF BIOLOGICAL SULPHATE REDUCTION

A well-tuned controller has parameters adapted to the dynamic properties of the specific process in order to have a fast and stable control system. If the process dynamic properties vary without the controller being re-tuned, the control system may become unstable or may become sluggish. The problems encountered with time varying process dynamics can be solved by tuning the controller in the most critical operation (conservative tuning) to guarantee the stability of the control when the process operates in a different operation point. However, if the tuning is too conservative the tracking speed is reduced, giving more sluggish control (Dunn et al., 2005). Another option to solving these problems is to use adaptive tuning, in which the controller parameters are varied along with variations of the process dynamics, so that the performance of the control system is maintained or optimized at any operation point. In such cases, a model for the process is necessary if the controller is to adapt to changing conditions. Typically, adaptive optimal control seeks for a maximum in a performance index function (Heinzle et al., 1993) and requires a model that accounts for changing process conditions (Figure 2.13; section 2.2).

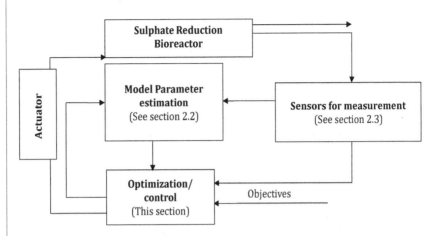

FIGURE 2.13 Adaptive control diagram for biological processes in bioreactors.

To our knowledge, there are still no studies reported on adaptive control for biological sulphate reduction systems. However, there are reports o-n adaptive control strategies designed for anaerobic digestion methanogenic processes. A few examples of these successful applications of adaptive control to anaerobic digestion processes are given below. These may be used as a starting point in the development of strategies to control biological sulphate reduction processes. The feed flow in an anaerobic digestion process was controlled in order to maintain the optimum production of methane and organic acids (Heinzle et al., 1993). A mechanistic simulation model appeared to be very useful in the design and testing of the adaptive controller. Once the parameters in the controller were determined, it was tested and an acceptable control of the one-stage bioreactor was obtained. The decrease of the flow rate by the controller corresponded to keeping the OLR of the reactor constant. Similarly, Steyer et al. (1999) designed a control strategy using the feed flow as the manipulated variable and biogas flow rate and pH as controlled variables. This control strategy, which was based on rather simple and reliable sensors that are widely used in industrial applications, was able to automatically monitor an anaerobic digestion process and to prevent overloadings. In addition, it was capable to adapt its parameters to any change in the influent concentration and thus to force the process to reach its maximum treatment ability.

Adaptive feedback control has been shown experimentally to be an appropriate tool to compensate for model uncertainties. Mailleret et al. (2004) proposed such a controller for a single substrate/single biomass model under general assumptions about the growth rate functions using the substrate concentration as a set point. Similarly, Dimitrova et al. (2011) aimed at stabilizing a four-dimensional nonlinear dynamic system, modelling the anaerobic degradation of organic wastes with methane production. A nonlinear feedback adaptive controller was proposed under general assumptions on the growth rate functions. The latter was able to stabilize asymptotically the dynamic system towards the unknown optimal (maximal) methane production.

In a different work, the acidity excess was avoided by maintaining a constant OLR in laboratory scale upflow anaerobic filters for the treatment of dairy and coffee effluent

(Johnson et al., 1995). Turbidity and conductivity were shown to be able to represent dissolved and suspended organic load. Thus, an adaptive feedback control model, based on the on-line determination of COD and gas production, was used to automatically vary the influent pumping rate and so avoid any imbalances and instability.

When systems involve highly complex and not fully understood processes, it becomes difficult to develop a mechanistic model which describes the system to its full extent. If the model cannot predict the effects of short timescale events, adaptive control will become less efficient. In these cases, utilizing black-box models such as artificial neural networks (ANN) are a suitable alternative (Zupan and Gasteiger, 1993; Zupan and Gasteiger, 1999; Holubar et al., 2002 and references therein). The latter do not require prior knowledge about the structure and relationships that exist between important variables. In addition, they are adaptable to system changes, in a short time scale, due to their learning abilities (Zupan and Gasteiger, 1999).

ANN have been applied with some success to anaerobic digestion systems (Wilcox et al., 1994; Holubar et al., 2002; Strik et al., 2005). Strik et al. (2005) successfully modelled the concentration of H_2S and NH_3 in biogas with ANN for the anaerobic digestion of flour Type W480 and peptone from casein (25:1) in a CSTR. The authors concluded that the developed ANN was suitable in predictive control tools. Atasoy et al. (2013) predicted the performance of fluidized bed reactors treating acid mine drainage using a designed, trained and validated ANN. Feed and effluent pH, feed sulphate, metal, COD concentration as well as operation time were used as input parameters. As output parameters, the effluent sulphate, COD, alkalinity and sulphide concentrations were taken. The ANN gave good agreement with experimental data showing the possibility to model complex systems without fully understanding the interactions within the various microbial groups present. Thus, ANN are also an attractive option to be applied to foresee and control the production of H_2S during various anaerobic processes.

TABLE 2.11 Control strategies for sulphate reduction processes.

Controlled variable	Manipulated variable	Controller type	System	Reference
Chemical sulphide concentration	Sulphide buffer flow	PI	Selective heavy metal precipitation (Cu-Zn and Pb-Zn)	Veeken et al. (2003)
Biogenic sulphide concentration	Sulphide buffer flow	PI	Zinc precipitation	König et al. (2006)
Chemical and biogenic sulphide concentration	Sulphide buffer flow	PI	Zinc precipitation	Esposito et al.(2006)
Chemical sulphide concentration	Sulphide buffer flow	PI	Copper and Zinc selective precipitation	Sampaio et al. (2009)
Chemical sulphide concentration	Sulphide buffer flow	PI	Zinc and Nickel selective precipitation	Sampaio et al. (2010)
Biogenic sulphide production	Organic loading rate (change in lactate concentration or in the HRT)	PID	Sulphate reduction in an Inversed fluidized bed reactor	Villa-Gomez et al. (2013)

2.4.4 EVALUATION ON CONTROL OF ANAEROBIC SULPHATE REDUCTION PROCESSES

For the full scale application of sulphur cycle based biotechnologies, it is of crucial importance to design and implement efficient control strategies to optimize the microorganisms growth and competition, to control inhibitory factors and/or to optimize the production of products for secondary processes, e.g., heavy metal precipitation with sulphide.

The number of full-scale applications of biocontrol approaches in sulphate reducing bioreactors is, however, still scarce. Work has been done on the comparison of different strategies to manipulate the production of biological sulphide (Villa-Gomez et al., 2014), which should be the starting point for the development of a control strategy. The latter should be accustomed to each specific case and adapted to the dynamic conditions of the biological sulphate reduction process by using adaptive control. One of the biggest obstacles in modelling and the design of control strategies relies on how to monitor the process variables. One way to overcome this gap may be to develop control strategies based on simple and available online measurements and on general assumptions of the processes or the use of so called software sensors (Keesman, 2002). The use of a complex model, i.e., with simplifications to a minimum, which can simulate well the dynamic processes in the bioreactor, is advisable to design and analyse control strategies before implementation as it can reduce significantly the experimentation time (Heinzle et al., 1993).

2.5 CONCLUSION

The development and application of a strategy for automated control of sulphate reduction bioprocesses in bioreactors is not an easy process as it should comprise and understand the complex dynamics of the process at hand. To attain a high controller performance for anaerobic biological processes, such as sulphate reduction, it is of crucial importance to develop models capable of simulating the chemical, physical and biological processes prevailing in the bioreactor and to correctly choose a sensor for online monitoring of the critical variables. There are very few reports on application of online sensors on biological sulphate reducing bioreactors. Thus, more research on sensor development and application is highly recommended as sensors are probably the biggest bottleneck when developing an automated sulphate reduction process.

In summary, to achieve automated sulphate reduction the following steps must be followed: 1) Define the control variables, e.g., sulphide production, microbial competition control, inhibition minimization; 2) Choose appropriate sensors, with additional software sensors, for online monitoring for substrates, products and/or intermediates; 3) Develop or adapt model to current needs with complexity level depending on its purpose (high complexity for developing control strategies or low complexity for control application) and 4) Combine the previous steps to develop an adaptive control strategy (examples given in this review). With the on-going research in modelling and in the development of sensors shown in this review, it becomes more attainable to design efficient control strategies for the automated biological sulphate reduction processes.

REFERENCES

Ai SY, Gao MN, Yang Y, Li JQ, Jin LT (2004a) Electrocatalytic Sensor for the Determination of Chemical Oxygen Demand Using a Lead Dioxide Modified Electrode. Electroanal 16: 404-409.

Ai SY, Li JQ, Ya Y, Gao MN, Pan ZS, Jin LT (2004b) Study on photocatalytic oxidation for determination of chemical oxygen demand using a nano-TiO_2–$K_2Cr_2O_7$ system. Anal Chim Acta 509: 237-241.

Andrews JF (1968) A mathematical model for the continuous culture of microorganisms utilizing inhibitory substance. Biotechnol and Bioeng 10: 707-723.

APHA (1995) Standard Methods for the Examinations of Water and Wastewater, 19th ed. American Public Health Association, Washington, DC.

Atasoy AD, Babar B, Sahinkaya E (2013) Artificial neural network prediction of the performance of upflow and downflow fluidized bed reactos treating acidic mine drainage water. Mine water environ 32:222-228.

Batstone DJ, Keller J, Angelidaki I, Kalyuzhnyi S, Pavlostathis SG, Rozzi A, Sanders W, Siegrist H, Vavilin V (2002) (IWA Task Group on Modelling of Anaerobic Digestion Processes). Anaerobic Digestion Model No. 1 (ADM1). IWA Publishing, London.

Berrocal MJ, Cruz A, Badr IHA, Bachas LG (2000) Tripodal ionophore with sulfate recognition properties for anion-selective electrodes. Anal Chem 72: 5295-5299.

Beyenal H, Sani RK, Peyton BM, Dohnalkova AC, Amonette JE, Lewandowski Z (2004) Uranium immobilization by sulfate-reducing biofilms. Environ Sci Technol 38: 2067-2074.

Bogomolova A, Komarova E, Reber K, Gerasimov T, Yavuz O, Bhatt S, Aldissi M (2009) Challenges of electrochemical impedance spectroscopy in protein biosensing. Anal Chem 81: 3944-3949.

Bourgeois W, Burgess JE, Stuetz RM (2001) On-line monitoring of wastewater quality: a review. J Chem Technol Biotechnol 76: 337-348.

Celis LB, Villa-Gomez D, Solis AGA, Morales BOO, Flores ER (2008) Characterization of sulfate reducing bacteria dominated surface communities during start-up of a down flow fluidized bed reactor. J Ind Microbiol Biotechnol 36: 111-121.

Chen JS, Zhang JD, Xian YZ, Ying XY, Liu MC, Jin LT (2005) Preparation and application of TiO_2 photocatalytic sensor for chemical oxygen demand determination in water research. Water Res 39: 1340-1346.

Cho JH, Kim YW, Na KJ, Jeon, GJ (2008) Wireless electronic nose system for real-time quantitative analysis of gas mixtures using micro-gas sensor array and neuro-fuzzy network. Sens Actuators B 134: 104-111.

Cord-Ruwisch R (1985) A quick method for the determination of dissolved and precipitated sulfides in cultures of sulfate-reducing bacteria. J Microbiol Methods 4: 33-36.

D'Acunto B, Esposito G, Frunzo L, Pirozzi F (2011) Dynamic modeling of sulfate reducing biofilms. Computers and Mathematics with Applications 62: 2601-2608.

Dimitrova N, Krastanov M (2011) Nonlinear adaptive stabilizing control of an anaerobic digestion model with unknown kinetics. Int J Robust Nonlinear Control 22: 1743-1752.

Dunn IJ, Heinzle E, Ingham J, Prenosil JE (2003) Biological reaction engineering: Dynamic modelling fundamentals with simulation examples. Wiley-VCH Verlag GmbH &Co. KgaA, Weinheim.

Dunn IJ, Heinzle E, Ingham J, Přenosil JE (2005) Automatic bioprocess control fundamentals In:. Biological Reaction Engineering. Wiley-VCH Verlag GmbH & Co. KGaA.

Esposito G, Weijma J, Pirozzi F, Lens PNL (2003) Effect of sludge retention time on H_2 utilization in a sulfate reducing gas-lift reactor. Process Biochem 39: 491-498.

Esposito G, Veeken A, Weijma J, Lens PNL (2006) Use of biogenic sulfide for ZnS precipitation. Sep Purif Technol 51: 31-39.

Esposito G, Lens P, Pirozzi F (2009) User-friendly mathematical model for the design of sulphate reducing H_2/CO_2 fed bioreators. J Environ Eng 135: 167-175.

Fedorovich V, Lens P, Kalyuzhnyi S (2003) Extension of Anaerobic Digestion Model No. 1 with processes of sulfate reduction. App Biochem and Biotechnol 109: 33-45.

Fibbioli M, Berger M, Schmidtchen FP, Pretsch E (2000) Polymeric membrane electrodes for monohydrogen phosphate and sulfate. Anal Chem 72: 156–160.

Firouzabadi S, Razavipanah I, Zhiani R, Ghanei-Motlagh M, Salavati MR (2013) Sulfate-selective electrode based on a bis-thiourea ionophore. Monatsh Chem 144: 113-120.

Fomichev AO, Vavilin VA (1997) The reduced model of self-oscillating dynamics in an anaerobic system with sulfate-reduction. Ecol Modell 95: 133-144.

Frevert T, Galster H (1978) Schnelle und einfache Methode zur In-Situ-Bestimmung von Schwefelwasserstoff in Gewässern und Sedimenten. Schweiz Z Hydrol 40: 199-208.

Frunzo L, Esposito G, Pirozzi F, Lens P (2012) Dynamic mathematical modeling of sulfate reducing gas-lift reactors. Process biochem 47: 2172-2181.

Ganjali MR, Naji L, Poursaberi T, Taghizadeh M, Pirelahi H, Yousefi M, Yeganeh-Faal A, Shamsipur M (2002) Novel sulfate ion-selective polymeric membrane electrode based on a derivative of pyrilium perchlorate. Talanta 58: 359-366.

Ganjali MR, Poujavid MR, Shamsipur M, Poursaeri T, Rezapour M, Javanbakht M, Sharghi H (2003) Novel Membrane Potentiometric sulfate ion sensor based on zinc-phthalocyanine for the quick determination of trace amounts of sulfate. Anal Sci 19: 995-999.

Ganjali MR, Ghorbani M, Daftari A, Norouzi P, Pirelahi H, Dargahani HD (2004) Highly selective liquid membrane sensor based on 1,3,5-Triphenylpyrylium perchlorate for quick monitoring of sulfate ions. Bull Korean Chem Soc 25: 172-176.

Gonzalez-Silva BM, Briones-Gallardo R, Razo-Flores E, Celis LB (2009) Inhibition of sulfate reduction by iron, cadmium and sulfide in granular sludge. J Hazard Mater 172: 400-407.

Grootscholten T, Keesman K, Lens P (2008) Modelling and on-line estimation of zinc sulphide precipitation in a continuously stirred tank reactor. Sep Purif Technol 3: 654-660.

Gupta A, Flora JRV, Sayles GD, Suidan MT (1994a) Methanogenesis and sulfate reduction in chemostats- II. Model development and verification. Water Res 28: 795-803.

Gupta A, Flora JRV, Gupta M, Sayles GD, Suidan MT (1994b) Methanogenesis and sulfate reduction in chemostats- I. Kinetic studies and experiments. Water Res 28: 781-793.

Guterman H, Ben-Yaakov S, Abeliovich A (1983) Determination of Total Dissolved Sulfide in the pH Range 7.5 to 11.5 by Ion Selective Electrodes. Anal Chem 55: 1731-1734.

Heinzle E, Dunn IJ, Ryhiner GB (1993) Modeling and control for anaerobic wastewater treatment. Adv Biochem Eng/Biotechnol 48: 79-114.

Holubar P, Zani L, Hager M, Fröschl W, Radak Z, Braun R (2002) Advanced controlling of anaerobic digestion by means of hierarchical neural networks. Water Res 36: 2582-2588.

Hu Y, Hua S, Li F, Jiang Y, Bai X, Li D, Niu L (2011) Green-synthesized gold nanoparticles decorated graphene sheets for label-free electrochemical impedance DNA hybridization biosensing. Biosens Bioelectron 26: 4355–4361.

Hübert T, Boon-Brett L, Black G, Banach U (2011) Hydrogen sensors - A review. Sens Actuators B 157: 329-352.

Huck C, Poghossian A, Wagner P, Schöning MJ (2012) Combined amperometric/field-effect sensor for the detection of dissolved hydrogen. Sens Actuators B 187: 168-173.

Jagadeesh CAP, Sudhaker RDS (2010) Modelling, Simulation and Control of Bioreactors Process Parameters Remote Experimentation Approach. Int J Comput Appl 1: 81–88.

Janssen PHM, Heuberger PSC (1995) Calibration of process-oriented models. Ecol Modell 83: 55–66.

Janssen AJH, Meijer S, Bontsema J, Lettinga G (1998) Application of the redox potential for controling a sulfide oxidizing bioreactor. Biotechnol Bioeng 60: 147-155.

Johnson KA, Wheatley AD, Fell CJ (1995) An application of adaptive control algorithm for the anaerobic treatment of an industrial effluent. IChemE 73: 203-211.

Jones BD, Ingle Jr JD (2005) Evaluation of redox indicators for determining sulfate-reducing and dechlorinating conditions. Water Res 39: 4343-4354.

Kaksonen AH, Puhakka JA (2007) Sulfate reduction bioprocesses for the treatment of acid mine drainage and the recovery of metals. Eng Life Sci 7: 541-564.

Kalyuzhnyi S, Fedorovich V (1997) Integrated mathematical model of UASB reactor for competition between sulphate reduction and methanogenesis. Water Sci Technol 36: 201-208.

Kalyuzhnyi S, Fedorovich V (1998) Mathematical modelling of competition between sulphate reduction and methanogenesis in anaerobic reactors. Bioresour Technol 65: 227-242.

Kalyuzhnyi S, Fedorovich V, Lens P, Hulshoff Pol LW, Lettinga G (1998) Mathematical modelling as a tool to study population dynamics between sulfate reducing and methanogenic bacteria. Biodegradation 9: 187-199.

Keesman KJ (2011) System Identification: an Introduction. Springer Verlag, UK.

Kim YC, Lee KH, Sasaki S, Hashimoto K, Ikebukuro K, Karube I (2000) Photocatalytic Sensor for Chemical Oxygen Demand Determination Based on Oxygen Electrode. Anal Chem 72: 3379-3382.

Klok JBM, van den Bosch PLF, Buisman CJN, Stams AJM, Keesman KJ, Janssen AJH (2012) Pathways of sulphide oxidation by haloalkaliphilic bacteria in limited-oxygen gas lift bioreactors. Environ Sci Technol 46: 7581-7586.

Klok JBM, de Graaff M, van den Bosch PLF, Boelee NC, Keesman KJ, Janssen AJH (2013) A physiologically based kinetic model for bacterial sulphide oxidation. Water Res 47: 483-492.

Knobel AN, Lewis AE (2002) A mathematical model of a high sulphate wastewater anaerobic treatment system. Water Res 36: 257-265.

König J, Keesman KJ, Veeken A, Lens PNL (2006) Dynamic modelling and process control of ZnS precipitation. Sep Sci Technol 41: 1025-1042.

Koydon S (2004) Contribution of sulfate-reducing bacteria in soil to degradation and retention of COD and sulfate. Karlsruhe University, Germany.

Kühl M, Jørgensen BB (1992) Microsensor measurements of sulfate reduction and sulfide oxidation in compact microbial communities of aerobic biofilms. Appl Environ Microbiol 58: 1164-1174.

Kühl M, Steuckart C, Eickert G, Jeroschewski P (1998) A H_2S microsensor for profiling biofilms and sediments: application in an acidic lake sediment. Aquat Microb Ecol 15: 201-209.

La Belle JT, Gerlach JQ, Svarovsky S, Joshi L (2007) Label-Free Impedimetric Detection of Glycan–Lectin Interactions. Anal Chem 79: 6959–6964.

Lens P, Visser A, Janssen A, Hulshoff Pol L, Lettinga G (1998) Biotechnological treatment of sulfate rich wastewaters. Crit Rev Environ Sci Technol 28: 41-88.

Lens PNL, Vallero M, Esposito G, Zandvoort M (2002) Perspectives of sulfate reducing bioreactor in environmental biotechnology. Rev Environ Sci Biotechnol 1: 311-325.

Lewis AE (2010) Review of metal sulphide precipitation. Hydrometallurgy 104: 222-234.

Li ZQ, Liu GD, Yuan LM, Shen GL, Yu RQ (1999) Sulfate-selective PVC membrane electrodes based on a derivative of imidazole as a neutral carrier. Anal Chim Acta 382: 165–170.

Li JQ, Li LP, Zheng L, Xian YZ, Ai SY, Jin LT (2005) Amperometric determination of chemical oxygen demand with flow injection analysis using $F-PbO_2$ modified electrode. Anal Chim Acta 548: 199-204.

Li JQ, Zheng L, Li L, Shi GY, Xian YZ, Jin LT (2006) Ti/TiO_2 Electrode preparation using laser anneal and its application to determination of chemical oxygen demand. Electroanal 18: 1014-1018.

Liamleam W, Annachhatre AP (2007) Electron donors for biological sulfate reduction. Biotechnol Adv 25: 452-463.

Lin YH, Lee KK (2001) Verification of anaerobic biofilm model for phenol degradation with sulphate reduction. J Environ Eng 2: 119-125.

Lin YH, Wu CL (2011) Sensitivity analysis of phenol degradation with sulphate reduction under anaerobic conditions. Environ Model Assess 16: 213-225.

López-Pérez PA, Neria-González MI, Flores-Cotera LB, Aguilar-Lopé, R (2013) A mathematical model for cadmium removal using a sulphate reducing bacterium: *Desulfovibrio alaskensis* 6SR. Int. J. Environ. Res. 7: 501-512.

Lyew D, Sheppard J (2001) Use of conductivity to monitor the treatment of acid mine drainage by sulphate-reducing bacteria. Water Res 35: 2081-2086.

Mailleret L, Bernard O, Steyer JP (2004) Nonlinear adaptive control for bioreactors with unknown kinetics. Automatica 40: 1379-1385.

Mašić A, Bengtsson J, Christensson M (2010) Measuring and modelling the oxygen profile in a nitrifying moving bed biofilm reactor. Math Biosci 227: 1-11.

Mattei MR, D'Acunto B, Esposito, G, Frunzo L, Pirozzi F (2014) Mathematical modeling of competition and coexistence of sulfate-reducing bacteria, acetogens, and methanogens in multispecies biofilms. Desalination and Water Treatment 1-9.

Miloshova M, Baltes D, Bychkov E (2003) New chalcogenic glass chemical sensors for S^{2-} and dissolved H_2S monitoring. Water Sci Technol 47: 135-140.

Mizani F, Rajabi F (2014) Potentiometric Sensor for Determination of Sulfate Ions based on 2-Amino-6-(tbutyl)-4-(pyridin-2-yl)pyrimidine)(dichlorido)palladium(II). Anal Bioanal Electrochem 6: 206-218.

Moosa S, Nemati M, Harrison STL (2002) A kinetic study on anaerobic reduction of sulphate, Part I: Effect of sulphate concentration. Chem Eng Sci 57: 2773-2780.

Moosa S, Harrison STL (2006) Product inhibition by sulphide species on biological sulphate reduction for the treatment of acid mine drainage. Hydrometallurgy 83: 214-222.

Morigi M, Scavetta E, Berrettoni M, Giorgetti M, Tonelli D (2001) Sulfate-selective electrodes based on hydrotalcites. Anal Chim Acta 439: 265-272.

Nagpal S, Chuichulcherm S, Peeva L, Livingston A (2000) Microbial sulfate reduction in a liquid-solid fluidized bed reactor. Biotechnol Bioeng 70: 370-380.

Nezamzadeh-Ejhieh A, Afshari E (2012) Modification of a PVC-membrane electrode by surfactant modified clinoptilolite zeolite towards potentiometric determination of sulfide. Micropor and Mesopor Mat 153: 267-274.

Nezamzadeh-Ejhieh A, Esmaeilian A (2012) Application of surfactant modified zeolite carbon paste electrode (SMZ-CPE) towards potentiometric determination of sulphate. Micropor and Mesopor Mat 147: 302-309.

Nishizawa S, Bühlmann P, Xiao KP, Umezawa Y (1998) Application of bis-thiourea ionophore for an anion selective electrode with a remarkable sulfate selectivity. Anal Chim Acta 358: 35–44.

Okabe S, Nielsen PH, Jones WL, Characklis WG (1995) Sulfide product inhibition of *Desulfovibrio desulfuricans* in batch and continuous cultures. Water Res 29: 571-578.

Oliveira MD, Correia MT, Coelho LC, Diniz FB (2008) Electrochemical evaluation of lectin-sugar interaction on gold electrode modified with colloidal gold and polyvinyl butyral. Colloids Surf B 66: 13–19.

Omil F, Lens P, Visser A, Hulshoff Pol LW, Lettinga G (1998) Long-term competition between sulfate reducing and methanogenic bacteria in UASB reactors treating volatile fatty acids. Biotechnol Bioeng 57: 676-685.

Overmeire A, Lens P, Verstraete W (1994) Mass transfer limitation of sulfate in methanogenic aggregates. Biotechnol Bioeng 44: 387-391.

Oyekola OO, Harrison STL, van Hille RP (2012) Effects of culture conditions on the competitive interaction between lactate oxidizers and fermenters in a biological sulphate reduction system. Bioresource Technology 104: 616-621.

Pandey SH, Kim KH, Tang KT (2012) A review of sensor-based methods for monitoring hydrogen sulfide. Trends Anal Chem 32: 87-99.

Parkin GF, Lynch NA, Kuo WC, Van Keuren EL, Bhattacharya SK (1990) Interaction between sulfate reducers and methanogens fed acetate and propionate. J Water Pollut Control Fed 62: 780-788.

Pind PF, Angelidaki I, Ahring BK, Stamatelatou K, Lyberatos G (2003) Monitoring and control of anaerobic reactors. Adv Biochem Eng/Biotechnol 82: 135-182.

Poinapen J, Ekama GA (2010) Biological sulphate reduction with primary sewage sludge in an upflow anaerobic sludge bed reactor – Part 6: development of a kinetic model for BSR. Water SA 36: 93-202.

Postgate JR (1984) The sulphate-reducing bacteria (2nd ed.). Cambridge: Cambridge University Press.

Qi P, Wan Y, Zhang D (2013a) Impedimetric biosensor based on cell-mediated bioimprinted films for bacterial detection. Biosens Bioelectron 39: 282-288.

Qi P, Zhang D, Wan Y (2013b) Determination of sulfate-reducing bacteria with chemical conversion from ZnO nanorods arrays to ZnS arrays. Sens Actuators B 181: 274-279.

Qu X, Tian M, Chen S, Liao B, Chen A (2010) Determination of chemical oxygen demand based on novel photoelectro-bifunctional electrodes. Electroanal 23: 1267-1275.

Ramsing NB, Küll M, Jørgensen BB (1993) Distribution of sulfate-reducing bacteria, O_2, and H_2S in photosynthetic biofilms determined by oligonucleotide probes and microelectrodes. Appl Environ Microbiol 59: 3840-3849.

Reis MAM, Lemos PC, Almeida JS, Carrondo MJT (1990) Influence of produced acetic acid on growth of sulfate reducing bacteria. Biotechnol Lett 12: 145-148.

Reis MAM, Almeida JS, Lemos PC, Carrondo MJT (1992) Effect of hydrogen sulfide on growth of sulfate reducing bacteria. Biotechnol Bioeng 40: 593-600.

Rezaei B, Khayamian T, Majidi N, Rahmani H (2009) Immobilization of specific monoclonal antibody on Au nanoparticles for hGH detection by electrochemical impedance spectroscopy. Biosens Bioelectron 25: 395–399.

Ribes J, Keesman K, Spanjers H (2004) Modelling anaerobic biomass growth kinetics with a substrate threshold concentration. Water Res 38: 4502-4510.

Rodriguez RP, Donoso-Bravo A, Valdiviesso GA, Torres I, Damasceno LHD, Zaiat, M (2011) Mathematical modelling of an horizontal-flow anaerobic immobilized biomass (HAIB) reactor treating acid mine drainage. X Latin American workshop and symposium on anaerobic digestion (DAAI), Ouro Preto, MG Proceedings.

Ryhiner GB, Heinzle E, Dunn IJ (1993) Modeling and Simulation of Anaerobic Wastewater Treatment and Its Application to Control Design: Case Whey. Biotechnol Prog 9: 332-343.

Sampaio RMM, Timmers RA, Xu Y, Keesman KJ, Lens PNL (2009) Selective precipitation of Cu from Zn in a pS controlled continuously stirred tank reactor. J Hazard Mater 165: 256-265.

Sampaio RMM, Timmers RA, Kocks N, André V, Duarte MT, van Hullebusch ED, Farges F, Lens PNL (2010) Zn–Ni sulfide selective precipitation: The role of supersaturation. Sep Purif Technol 74: 108-118.

Santegoeds CM, Ferdelman TG, Muyzer G, de Beer D (1998) Structural and functional dynamics of sulfate-reducing populations in bacterial biofilms. Appl Environ Microbiol 64: 3731-3739.

Santegoeds CM, Damgaard LR, Hesselink G, Zopfi J, Lens P, Muyzer G, de Beer D (1999) Distribution of sulfate-reducing and methanogenic bacteria in anaerobic aggregates determined by microsensor and molecular analysis. Appl Environ Microbiol 65: 4618-4629.

Sasaki S, Yokoyama K, Tamiya E, Karube I, Hayashi C, Arikawa Y, Numata M (1997) Sulfate sensor using *Thiobacillus ferrooxidans*. Anal Chim Acta 347: 275-280.

Sawyer CN, McCarty PL, Parkin GF (2003) Chemistry for environmental engineering and science. McGraw-Hill.

Schmidt E, Marton A, Hlavay J (1994) Determination of the total dissolved sulphide in the pH range 3-11.4 with sulphide selective ISE and Ag/Ag_2S electrodes. Talanta 41: 1219-1224.

Shamsipur M, Yousefi M, Hosseini M, Ganjali MR, Sharghi H, Naeimi H (2001) A Schiff base complex of Zn(II) as a neutral carrier for highly selective PVC membrane sensors for the sulfate ion. Anal Chem 73: 2869–2874.

Shamsipur M, Yousefi M, Ganjali MR, Poursaberi T, Faal-Rastgar M (2002) Highly selective PVC-membrane electrode based on 2,5-diphenyl-1,2,4,5-tetraaza-bicyclo[2.2.1]heptane as a neutral carrier. Sens Actuators B 82: 105-110.

Shin HS, Oh SE, Bae BU (1996) Competition between SRB and MPB according to temperature change in the anaerobic treatment of tannery wastes containing high sulfate. Environ Technol 17: 361-370.

Silva CR, Conceição CDC, Bonifácio VG, Filho OF, Teixeira MFS (2008) Determination of the chemical oxygen demand (COD) using a copper electrode: a clean alternative method. J Solid State Electrochem 13: 665-669.

Sötemann SW, van Rensburg P, Ristow NE, Wentzel MC, Loewenthal RE, Ekama GA (2005) Integrated chemical, physical and biological processes modelling Part 2 – Anaerobic digestion of sewage sludges. Water SA 3: 545-568.

Spanjers H, Weijma J, Abusam A (2002) Modelling the competition between sulphate reducers and methanogens in a thermophilic methanol-fed bioreactor. Water Sci Technol 45: 93-98.

Speece RE (1996) Anaerobic biotechnology for industrial wastewaters. Tennessee: Archae Press.

Stephanopoulos G (1984) Chemical Process Control: An Introduction to Theory and Practice. Prentice-Hall: Englewood Cliffs, NJ.

Steyer JP, Buffière P, Rolland D, Moletta R (1999) Advanced control of anaerobic digestion processes through disturbances monitoring. Water Res 33: 2059-2068.

Strik DPBTB, Domnanovich AM, Zani L, Braun R, Holubar P (2005) Prediction of trace compounds in biogas from anaerobic digestion using the MATLAB Neural Network Toolbox. Environ Modell Softw 20: 803-810.

Tang Y, Ontiveros-Valencia A, Feng L, Zhou C, Krajmalnik-Brown R, Rittman BF (2012) A biofilm model to understand the onset of sulfate reduction in denitrifying membrane biofilm reactors. Biotechnol Bioeng 110: 763-772.

Thót I, Solymosi P (1988) Szabó Z. Application of a sulphide-selective electrode in the absence of a pH-buffer. Talanta 35: 783-788.

Torner-Morales FJ, Buitrón G (2010) Kinetic characterization and modelling simplification of an anaerobic sulfate reducing batch process. J Chem Technol Biotechnol 85: 453-459.

Vavilin VA, Vasiliev VB, Ponomarev AV, Rytow SV (1993) Simulation model 'METHANE' as a tool for effective biogas production during anaerobic conversion of complex organic matter. Bioresour Technol 48: 1-8.

Vavilin VA, Vasiliev VB, Rytov SV, Ponomarev AV (1994) Self-oscillating coexistence of methanogens and sulfate-reducers under hydrogen sulfide inhibition and the pH-regulating effect. Bioresour Technol 49: 105-109.

Veeken AHM, de Vries S, van der Mark A, Rulkens WH (2003) Selective precipitation of heavy metals as controlled by a sulfide-selective electrode. Sep Sci Technol 38: 1-19.

Villa-Gomez DK, Cassidy J, Keesman K, Sampaio R, Lens PNL (2014) Sulfide response analysis for sulphide control using a pS electrode in sulphate reducing bioreactors. Water Res 50: 48-58.

Wan Y, Zhang D, Hou B (2009) Monitoring microbial populations of sulfate-reducing bacteria using a impedimetric immunosensor based on agglutination assay. Talanta 80: 218-223.

Wan Y, Zhang D, Wang Y, Hou B (2010a) A 3D-impedimetric immunosensor based on foam Ni for detection of sulfate-reducing bacteria. Electrochem Commun 12: 288–291.

Wan Y, Zhang D, Hou B (2010b) Determination of sulphate-reducing bacteria based on vancomycin-functionalised magnetic nanoparticles using a modified quartz crystal microbalance. Biosens Bioelectron 25: 1847-1850.

Wan Y, Zhang D, Hou B (2010c) Selective and specific detection of sulfate-reducing bacteria using potentiometric stripping analysis. Talanta 82: 1609-1611.

Wan Y, Lin ZF, Zhang D, Wang Y, Hou BR (2011a) Impedimetric immunosensor doped with reduced graphene sheets fabricated by controllable electrodeposition for the non-labelled detection of bacteria. Biosens Bioelectron 26: 1959–1964.

Wan Y, Zhang D, Wang Y, Qi P, Hou B (2011b) Direct immobilisation of antibodies on a bioinspired architecture as a sensing platform. Biosens Bioelectron 26: 2595–2600.

Wang J, Li K, Yang C, Wang Y, Jia J (2012) Ultrasound electrochemical determination of chemical oxygen demand using boron-doped diamond electrode. Electrochem commun 18: 51-54.

Wilcox SJ, Hawkes DL, Hawkes FR, Guwy AJ (1994) A neural network, based on bicarbonate monitoring, to control anaerobic digestion. Water Res 29: 1465-1470.

Xu Xj, Chen C, Wang Aj, Guo Hl, Yuan Y, Lee DJ, Ren, Nq (2014) Kinetics of nitrate and sulphate removal using a mixed microbial culture with or without limited-oxygen fed. Appl Microbiol Biotechnol 98: 6115:6124.

Yu H, Wang H, Quan X, Chen S, Zhang Y (2007) Amperometric determination of chemical oxygen demand using boron-doped diamond (BDD) sensor. Electrochem Commun 9: 2280–2285.

Zhang SQ, Li LH, Zhao HJ (2009a) A portable photoelectrochemical probe for rapid determination of chemical oxygen demand in wastewaters. Environ Sci Technol 43: 7810-7815.

Zhang J, Zhou B, Zheng Q, Li J, Bai J, Liu Y, Cai W (2009b) Photoelectrocatalytic COD determination method using highly ordered TiO_2 nanotube array. Water Res 43: 1986-1992.

Zhao HJ, Jiang DL, Zhang SQ, Catterall K, John R (2004) Development of a direct photoelectrochemical method for determination of chemical oxygen demand. Anal Chem 76: 155-160.

Zheng Q, Zhou BX, Bai J, Li LH, Jin ZJ, Zhang JL, Li JH, Liu YB, Cai WM, Zhu XY (2008) Self-organized TiO_2 nanotube array sensor for the determination of chemical oxygen demand. Adv Mater 20: 1044-1049.

Zhu LH, Chen Y, Wu YH, Li XR, Tang HQ (2006) A surface-fluorinated-TiO_2–$KMnO_4$ photocatalytic system for determination of chemical oxygen demand. Anal Chim Acta 571: 242-247.

Zupan J, Gasteiger J (1993) Neural Networks for chemists. Weinheim:VCH.

3

Sulphide Response Analysis for Sulphide Control using a pS Electrode in Sulphate Reducing Bioreactors

CHAPTER 3

Sulphide Response
Analysis for
Sulphide Control
using a pS
Electrode in
Sulphate Reducing
Bioreactors

ABSTRACT

Step changes in the organic loading rate (OLR) through variations in the influent chemical oxygen demand (COD_{in}) concentration or in the hydraulic retention time (HRT) at constant COD/SO_4^{2-} ratio (0.67) were applied to create sulphide responses for the design of a sulphide control in sulphate reducing bioreactors. The sulphide was measured using a sulphide ion selective electrode (pS) and the values obtained were used to calculate proportional-integral-derivative (PID) controller parameters. The experiments were performed in an inverse fluidized bed bioreactor with automated operation using the LabVIEW software version 2009®. A rapid response and high sulphide increment was obtained through a stepwise increase in the COD_{in} concentration, while a stepwise decrease to the HRT exhibited a slower response with smaller sulphide increment. Irrespective of the way the OLR was decreased, the pS response showed a time-varying behavior due to sulphide accumulation (HRT change) or utilization of substrate sources that were not accounted for (COD_{in} change). The pS electrode response, however, showed to be informative for applications in sulphate reducing bioreactors. Nevertheless, the recorded pS values need to be corrected for pH variations and high sulphide concentrations (>200 mg/L).

This chapter has been published as:
Villa-Gomez DK, Cassidy J, Keesman KJ, Sampaio R, Lens PNL (2014) Sulfide response analysis for sulphide control using a pS electrode in sulphate reducing bioreactors. Water Res 50: 48–58.

3.1 INTRODUCTION

Biological sulphate reduction is a process for the treatment of metal containing wastewaters enabling the recovery of metals as sulphidic precipitates (Bijmans et al., 2011). Sulphate reducing bacteria (SRB) reduce sulphate through the oxidation of either organic compounds or hydrogen, resulting in the production of sulphide (Kaksonen and Puhakka, 2007). Most of the metal containing wastewaters are deficient in organic compounds (Papirio et al., 2013). Thus, their addition as electron donor for sulphate reduction determines the overall costs of the process (Gibert et al., 2004; Zagury et al., 2006). For metal removal and recovery processes, the required amount of sulphide to be produced by SRB depends on the composition of the wastewater to be treated, i.e. its metal concentration. Steering the sulphide production towards this required stoichiometric amount in bioreactors is highly relevant to avoid overproduction of H_2S that increases operational costs and may require a sulphide removal post-treatment step.

Process control has been used for several biological production processes yielding desirable end products such as ethanol, penicillin and diverse fermentation products as well as for wastewater treatment (Dunn et al., 2005). In these processes, typically, the set-point control is based on the manipulation of temperature, pH, substrate or dissolved oxygen concentration. In anaerobic digestion, control variables commonly used for process control are intermediate compounds such as volatile organic acids, pH, bicarbonate concentration, alkalinity or gas concentrations/flow rates (Pind et al., 2003).

Even though large progress has been made in the control of anaerobic (methanogenic) systems, there is insufficient knowledge about process control of sulphate reducing bioreactors. Mathematical models have been developed to support the design of a control strategy for sulphate reduction in bioreactors (Gupta et al., 1994; Kalyuzhnyi and Fedorovich, 1998; Oyekola et al., 2012). In these studies the objective was to outcompete or favor microbial trophic groups other than SRB, while accounting for the control of the sulphide production. Torner-Morales and Buitrón (2010) used the redox potential as a control variable to maintain the sulphate reduction efficiency and subsequent partial sulphide oxidation in a single sequencing batch reactor unit. The control of the redox potential, combined with the control of the pH, allowed a combined sulphate-reducing/sulphide-oxidizing process with a continuous operation and a significant yield of elemental sulfur (64%).

The development of a control strategy based on the sulphide concentration as the controlled variable is a more direct approach than the redox potential for the control of the sulphide concentration in sulphate reducing bioreactors. For this, the most adequate sensor developed is the pS electrode that measures the activity of the S^{2-} species (Veeken et al., 2003a). The use of a pS electrode in biological systems was already described by Dan et al., (1985) to monitor photosynthetic sulphide oxidation by *Chlorobium phaeobacteroides*, by Bang et al. (2000) to determine sulphide concentrations produced by expression of the thiosulphate reductase gene (phsABC) from *Salmonella enterica* Serovar Typhimurium in *Escherichia coli* for heavy metal removal and by Yamaguchi et al. (2001) to examine intra-granule sulphide profiles in anaerobic granular sludge. This sensor, in combination with a pH electrode, has also been successfully validated in the control of the sulphide concentration in precipitator reactors for selective metal recovery using chemically and biologically produced (biogenic) sulphide (Grootscholten et al., 2008; König et al., 2006, Sampaio et al., 2009; Veeken et al., 2003a; Veeken et al., 2003b).

The selection of an appropriate control strategy largely depends on the process characteristics. A proportional-integral (PI) control strategy using the pS electrode was found to be sufficient for biogenic sulphide control entering a precipitator reactor with only metal sulphide precipitation taking place (König et al., 2006). Controlling the sulphide concentration directly in the SRB bioreactor is more complex as it needs to take into account the biological sulphide production process as well. Therefore, it requires a control strategy that manipulates the organic loading rate (OLR) to the sulphide producing SRB bioreactor, where the set-point ideally depends on the metal concentration that is required to be precipitated. Thus, an additional control parameter is required to overcome the lag time between substrate dosing, substrate bioconversion and release of the desired product (S^{2-}). Even though robust control strategies are promising options over conventional control types such as the PI or the proportional-integral-derivative (PID) controller, there are only a few experimental results in the literature validating their application in anaerobic bioreactors (Steyer et al., 1999). In contrast, the PID controller has been widely used in anaerobic bioreactors (Dunn et al., 2003; Dunn et al., 2005; Jagadeesh and Sudhaker 2010; Marsili-Libelli and Beni 1996; Pind et al., 2003) and contains, in addition to the parameters of the PI control, a derivative control parameter to overcome lag phases. Therefore, the PID controller can be considered as a valid option to control the sulphide concentration in SRB bioreactors. The PID control parameters can be obtained by using different tuning strategies and tested experimentally or through model simulations (Pind et al., 2003).

The aim of this study was to evaluate strategies to manipulate the OLR to control the sulphide concentration in an inverse fluidized bed (IFB) bioreactor using a pS electrode. The OLR was manipulated via changing the influent COD (COD_{in}) concentration or the hydraulic retention time (HRT). The evaluation of the strategies was based on the analysis of the response of the system to the applied change in terms of response time and time delay, load and set point (sulphide) changes, as well as robustness and stability of the sensor (Rodrigo et al., 1999). The Cohen-Coon method (Dunn et al., 2005) was used to determine the parameters of the PID controller. Since pS sensors have not yet been applied in sulphate reducing bioreactors, their feasibility for application was also assessed.

3.2 MATERIALS AND METHODS

3.2.1 REACTOR SET-UP

The experiments were carried out in an IFB bioreactor as described by Villa-Gomez et al. (2014), but with automated operation using a data acquisition card (NI cDAQ-9174, National Instruments, The Netherlands) and Labview software version 2009® (Figure 3.1). Lactate was used as electron donor and carbon source and sulphate was added as Na_2SO_4 at a COD/ SO_4^{2-} ratio of 0.67. The synthetic medium used in the bioreactor experiment was the same as used by Villa-Gomez et al. (2014).

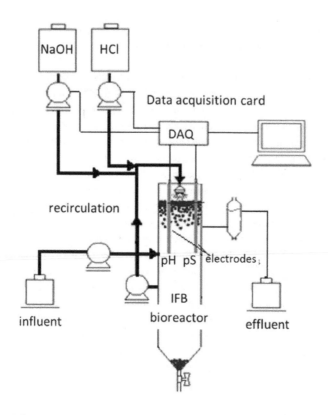

FIGURE 3.1 Experimental set-up of the IFB bioreactor with pH and pS online measurement as well as pH control.

The pH and pS in the IFB bioreactor were monitored using a sulphide resistant pH electrode (Prosense, Oosterhout, The Netherlands) and a solid state Ag_2S ion selective (S^{2-}) electrode (Prosense, Oosterhout, The Netherlands) of 40 cm length, each inserted on the top of the bioreactor column (Figure 3.) with a lower and upper detection limit of 0.003 and 3200 mg/L, respectively, a response time <10 seconds after first immersion in the sample and a sensitivity of -26 ± 3 mV/decade (Prosense Oosterhout, The Netherlands). The Labview software® contained a PID controller (PID and Fuzzy Logic Toolkit, National Instruments, The Netherlands) for the control of the pH using stock solutions of HCl/NaOH connected to the recirculation tube and with PID parameters obtained by error minimization.

Prior to this study, the IFB bioreactor was operated for over 150 days. At the beginning of this study, the biofilm consisted predominantly of incompletely oxidizing SRB, fed with lactate (16.7 gCOD/gVSS.d) and producing acetate and sulphide as final products. Acetate utilization by SRB also occurred in the bioreactor (37.1 gCOD/gVSS.d). Methane producing activity was only observed at bioreactor start-up (22.2 gCOD/gSSV.d), while no methane production was detected anymore when this study began.

3.2.2 EXPERIMENTAL DESIGN

Several OLR step changes were applied in the IFB bioreactor to create responses in terms of the sulphide concentration produced by SRB, in order to determine the coefficients of the PID controller. The experiments consisted of running the IFB bioreactor at a constant OLR and, once the pS electrode displayed constant values for at least 5 HRTs, to change this OLR to create a step response in the sulphide concentration until a new steady state was reached in the pS response (response time). The OLR was changed from 0.5 to 1 gCOD/L.d through either variations in the COD_{in} concentration or in the HRT by changing the influent flow rate. Once the reactor reached a new steady state in the pS values, the same methodology was applied to return the OLR to its initial setting.

The reliability of the pS electrode response under the IFB bioreactor operational conditions was also evaluated. The pS electrode response in the IFB bioreactor was analyzed after variations in the total dissolved sulphide (TDS) concentrations ranging from 100 to 500 mg/L. In addition, the pS signal was analyzed at TDS concentrations ranging from 20 mg/L to 400 mg/L at constant pH of 7 with chemically produced sulphide (Na_2S) and with biogenic sulphide. Chemically produced sulphide stock solutions were prepared with Na_2S*H_2O (Merck, extra pure, about 35% Na_2S), while biogenic sulphide samples were taken from the IFB bioreactor at different operation periods and thus, different TDS concentrations. In the chemically produced samples, TDS concentrations were re-measured once the pH was adjusted to 7, as the sulphide concentration is directly related to the pH. In the samples taken from the bioreactor liquid, the pH was not adjusted as the bioreactor was already at pH 7.0 ($\pm$ 0.2). The theoretical S^{2-} was calculated using equation 3.1; these values were converted to mV with the calibration line made following the pS electrode calibration procedure (section 2.3).

$$S^{2-} = \frac{TDS}{1 + \frac{(H^+)}{K_{a2}} + \frac{(H^+)^2}{K_{a2}K_{a1}}} \qquad (3.1)$$

Where: is measured by the pS electrode; $K_{a1} = 10^{-7}$ and $K_{a2} = 10^{-13.9}$

The pS electrode measures the concentration of the S^{2-} species (pS= -log [S^{2-}]), which depends on the TDS and pH (equation 3.1). Hence, an increase in the TDS leads to a decrease of the pS values. For practicality, sulphide in this manuscript refers to all sulphide species (HS⁻, S^{2-} and H_2S), while TDS (all dissolved sulphide species) and pS (S^{2-}) are specific determinations of the sulphide species.

3.2.3 pS ELECTRODE CALIBRATION

The pS electrode (Prosense, Oosterhout, The Netherlands) operates with a silver/sulphide sensing element, which in contact with a solution containing sulphide ions develops an electrode potential that is measured against a constant reference potential. This measured potential is described by the Nernst equation (Gründig and Krabisch, 1989) which is a linear function of the logarithm of the activity of sulphide (S^{2-}):

$$E = E_o + b * log (S^{2-}) \qquad (3.2)$$

The calibration of the pS electrode followed the methodology and principles described by

Veeken et al. (2003b) and Sampaio et al. (2009). However, some modifications were made considering the variation of the pS electrode readings due to the biotic conditions in this study.

Approximately 10 mM (320 mg/L) of Na_2S were titrated with 1 M of HCl, from high (12) to low (2.5) pH values in a solution containing also the synthetic medium used in the bioreactor influent. This was done to consider its influence on the ionic strength, which varies the activity coefficient of each ion in the solution (Gründig and Krabisch, 1989) and thus, also the pS values. VisuaVisualMINTEQ version 3.0 (US EPA, 1999, http://www.lwr.kth.se/English/ OurSoftware/vminteq/index.html) was used to calculate the ionic strength applying the Davies equation. The values obtained for the calibration curve from the chemical sulphide solution and the bioreactor media (including the contribution of lactate and acetate anions) did not display important differences (0.07 ± 0.005) and thus, the extrapolation of the pS electrode values obtained with the calibration curve were considered reliable to calculate the pS values obtained in the bioreactor. It is also important to mention that no studies were found reporting pS electrode interferences due to the anaerobic conditions and, in fact, sulphide measurements with pS electrodes are preferably carried out under anaerobic conditions to avoid sulphide variations due to contact with air that allows the formation of other sulfur species (Brown et al., 2011).

Two pS electrodes were used during this study: the first electrode exhibited a calibration curve that covered a range in voltage from 119 mV to 470 mV (Figure 3.2), while the working range of the second electrode covered from 353 to 672 mV (Figure 3.2). Both electrodes did not show significant variations in the repetitions of the calibration curve during this work and displayed a correlation coefficient > 0.99 when the voltage was related to the logarithm of the sulphide activity (Equation 3.2- Figure 3.2b). The slope values obtained with both pS electrodes are within the range of slope values for divalent anions (> 22 mV) in ion selective electrodes (Gründig and Krabisch, 1989). Additionally, other authors have obtained similar values with pS eletrodes following the same principle (Equations 3.1 and 3.2) (Sampaio et al., 2009; Veeken et al., 2003b).

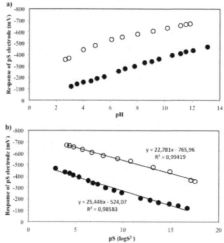

FIGURE 3.2 Response in mV of the two pS electrodes used in this study as a function of a) pH and b) pS.

3.2.4. PID CONTROLLER PARAMETERS

The pS electrode output values obtained from the step responses via either a change in the COD_{in} concentration or the HRT were used to determine the PID controller parameters. From these step responses, the following characteristics were determined: gain (K), time constant (τ) and time delay (td) (Equation 3.3):

$$K = \frac{output(at\,steady\,state)}{input(at\,steady\,state)} = \frac{B}{A}$$

$$\tau = \frac{B}{S}$$

(3.3)

$$t_d = \text{time elapsed until the system responded (time delay)}$$

where B is equal to $[pS_{final} - pS_{initial}]$, A is equal to $[OLR_{final} - OLR_{initial}]$ and S is the slope of the sigmoidal response at the point of inflection.

The Cohen-Coon tuning method (Dunn et al., 2005) was used to provide estimates of the PID parameters (K_C, τ_i, τ_D) on the basis of the experimental step response data (Equation 3.3):

$$K_c = \frac{1}{K}\frac{\tau}{t_d}\left(\frac{4}{3}+\frac{t_d}{4f}\right) \qquad \tau_i = t_d\,\frac{21+6t_d/\tau}{18+8t_d/\tau} \qquad \tau_D = t_d\,\frac{4}{11+2t_d/\tau} \qquad (3.4)$$

3.2.5 RESIDENCE TIME DISTRIBUTION IN THE IFB BIOREACTOR

To discard time delays in the sulphide response due to the hydrodynamic behavior of the system, the residence time distribution (RTD) of the IFB bioreactor was determined as described by Warfvinge (2009) at a theoretical HRT of 12 and 24 h. The residence time distribution curves were determined using LiCl, since this compound is not degraded nor adsorbed by microorganisms (Olivet et al,. 2005). LiCl was dissolved in small amounts of water and then injected at the top of the bioreactor column with a syringe over a time as short as possible. The amount of tracer used corresponded to a bulk concentration of 30 mg/L in the bioreactor working volume (2.5 L). Samples of the effluent were taken at predetermined time intervals until the recovery of the tracer was completed.

3.2.6 ANALYSES

COD was measured by the closed reflux method (APHA 2005). Sulphate was measured as described by Villa-Gomez et al. (2011). TDS was determined spectrophotometrically by the colorimetric method described by Cord-Ruwish (1985) using a spectrophotometer (Perkin Elmer Lambda 20). Acetate was measured by gas chromatography (GC-CP 9001 Chrompack) after acidification of the samples with 5% concentrated formic acid and filtration through a 0,45 µm nitrocellulose filter (Millipore). The gas chromatograph was fitted with a WCOT fused silica column, the injection and detector temperatures were 175 and 300 °C, respectively.

3.3. RESULTS

3.3.1. ANALYSIS OF THE PS ELECTRODE SIGNAL

Figure 3.3a shows a typical pS electrode response to a change in TDS concentration in the IFB bioreactor operation in the TDS concentration range between 100 and 500 mg/L. The pS response was consistent with the variations in TDS concentration. The small jumps in the signal on days 3, 6, 13 and 15 were due to small variations in pH (± 0.3). Variations in the TDS concentration above 200 mg/L did not induce noticeable changes in the output pS value compared to the TDS concentrations below 200 mg/L. A similar trend was found in the pS response at low and high sulphide concentrations when the pS signal was tested to check its reliability at constant pH for biogenic and chemically produced sulphide (Figure 3.3b and Figure 3.3c). The pS values show a logarithmic response with an exponential increase from approximately 0 to 100 mg/L of TDS (Figure 3.3b). Above this concentration, the pS values differed considerably between the biogenic and chemically produced sulphide as well as for the theoretical sulphide concentration (Figure 3.3c). This is because the differences in sulphide sampling between the sulphide sources (inside the bioreactor and in open bottles) are more noticeable at higher sulphide concentrations as the voltage response obtained from the S^{2-} concentration is less pronounced due to its logarithmic behavior (Equation 3.1) and thus, small variations in sulphide concentration highly affect the recorded pS values.

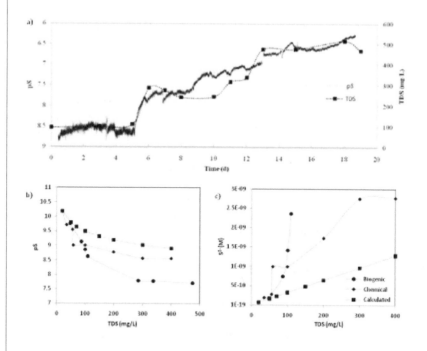

FIGURE 3.3 a) Typical response of the pS electrode to the sulphide production increment in the IFB bioreactor, b) Relationship between pS and TDS concentration, c) Relationship between S^{2-} (M) and TDS concentration.

3.3.2. RESPONSE OF THE BIOREACTOR TO CHANGES IN OLR

The step changes in the COD_{in} concentration from 0.5 to 1 and from 1 to 0.5 gCOD/L.d increased and decreased, respectively, the sulphate removal efficiency (Table 3.1), while the COD removal efficiency maintained similar values despite the OLR change (>80%). Prior to the first step response, no acetate could be detected, while the change in OLR to 1 and later to 0.5 gCOD/L.d caused an accumulation of acetate in the system to 28.2 and 52.2 mg/L, respectively. The TDS concentration increased from 113.4 to 333.1 mg/L with the change in OLR from 0.5 to 1 gCOD/L.d The return to the initial OLR conditions yielded a different TDS concentration value at the steady state (152.8 mg/L) as compared to the values prior the first step change (Table 3.1).

The step response from 0.5 to 1 gCOD/L.d through a change in the HRT caused an increase in both the TDS concentration (131 to 160.8 mg/L) and the sulphate removal efficiency (39 to 63.2%), while the return to the initial OLR value (0.5 gCOD/L.d) further increased both values (Table 3.1). The acetate concentration increased from 37.5 to 121.6 mg/L after the increase in the OLR, while upon return to the initial OLR, acetate values were similar to the ones obtained prior to the step responses. The COD removal efficiency did not show notice-able changes after the stepwise changes in the OLR and only a slight increase (75.9%) was observed once the OLR returned to its initial value.

TABLE 3.1 Steady state values in the IFB bioreactor prior to and after a change in COD_{in} concentration and HRT (± deviation error, negligible for pS and pH).

	Change in COD$_{in}$			Change in HRT		
OLR (gCOD/L.d)	0.5	1	0.5	0.5	1	0.5
COD (g/L)	0.5	1	0.5	0.5	0.5	0.5
HRT (h)	24	24	24	24	12	24
Mean residence time (h)	21.7	21.7	21.7	21.7	11.15	21.7
SO$_4^{2-}$ removal (%)	39.3 ± 6.3	56.4 ± 13.9	24.6 ± 6.5	39.0 ± 5.4	63.2 ± 3.5	76.9 ± 15.1
COD removal (%)	86.3 ± 6.1	78.6 ±14	83.4 ± 6.5	63.8 ± 13.9	59.3 ± 5.5	75.9 ± 0.4
Acetate (mg/L)	0	28.2 ± 7.7	52.2	37.5 ± 6.5	121.6 ±20.1	34.2 ± 13.2
TDS (mg/L)	113.4 ± 15.68	333.1 ± 58.1	152.8± 6.5	131.0 ± 6.31	160.8 ± 12.8	207 ± 21.4
pH	6.8	7	6.9	6.9	6.7	6.8
pS	8.76	7.6	8.05	11.07	9.97	9.32
pS (mV)	-567	-592.7	-582.5	-513.7	-538.8	-553.8

The mean residence time in the IFB bioreactor for the theoretical HRT of 24 and 12 h was 21.7 and 11.15 h, respectively (Table 3.1). The RTD analysis (Figure 3.4) showed a rapid increase in Li$^+$ concentration in time followed by a slow, steady tail, which resembles the hydrodynamic behavior of a completely mixed reactor at both HRTs studied (Warfvinge, 2009).

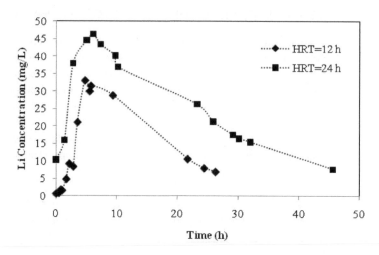

FIGURE 3.4 Residence time distribution profile in the IFB bioreactor at mean residence times of 12 and 24 h.

3.3.3. RESPONSE OF THE PS ELECTRODE TO CHANGES IN OLR

The step response via a change in the COD_{in} displayed an exponential increase in the sulphide concentration that lasted 4 days with no significant time delay and reached a steady state pS value of 7.6 after the OLR increment (Figure 3.5a). When the OLR was subsequently decreased to 0.5 gCOD/L.d, the pS response time was longer and the pS values at steady state (pS 8.0) were different compared to the ones at the start of the experiment. The time delay in the response of the system was less than 24 h and the decrease of the sulphide concentration lasted 2 days until the values reached a steady state.

The change in HRT showed a time delay of approximately 7 days and an exponential curve that lasted 15 days following the OLR step change (Figure 3.5b). The sulphide concentration slowly increased and consequently, the pS decreased from 11.1 to 9.9. Returning the HRT from 12 to 24 h showed a faster response in the system with a time delay of less than one day, reaching a steady state after 2 days. Surprisingly, the return to the initial conditions led to a further increase in the sulphide concentration giving a new pS value of 9.3 at steady state. During this experiment, strong drops in the pS values (sharp peaks) were observed on day 7 and 30 due to fouling of the ion selective membrane that reduced its sensitivity (Gründig and Krabisch, 1989). The electrode sensitivity was, however, recovered immediately as this fouling was non-adhesive and thus, the bioreactor liquor contact cleaned the membrane.

The pS signal showed in general negligible fluctuations in the signal for a change in the COD_{in}, while for a change in HRT, the signal fluctuated prior to the step change in OLR. This was due to small variations in the pH (Figure 3.5b), which induced changes in the S^{2-} species (Equation 3.1).

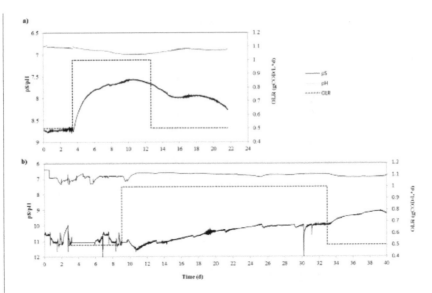

FIGURE 3.5 Step responses of the pS electrode response for a change in a) COD$_{in}$ and b) HRT.

3.3.4. COHEN-COON TUNING COEFFICIENTS AND PID PARAMETERS

The Cohen-Coon tuning coefficients (Equation 3.3-Table 3.2) and PID parameters (Equation 3.4-Table 3.2) were determined based on the analysis of the pS responses (Figure 3.5). Additionally, the response time and the ΔTDS are also included in Table 3.2. A step change in the OLR from 0.5 to 1 gCOD/L.d induced by a change in the COD$_{in}$ concentration showed the smallest response time with an important increment in the sulphide concentration (Table 3.2), while the step change induced by the variation in the HRT, exhibited a longer response time and a lower sulphide increment was observed for the same step change in OLR (Table 3.2).

The coefficients obtained from the pS responses showed that the sulphide load change (K) for an increment in the OLR is higher in contrast to the values obtained when the OLR is decreased for both OLR change strategies (Table 3.2). Furthermore, the value obtained for a decrease in the OLR via a change in the HRT was positive, in contrast with the other K parameters. This was due to the increase in the sulphide concentration, despite the decrease in the OLR that lowers the substrate available for sulphate reduction. τ values display differences that are due to the differences in slope (Equation 3.1) of the exponential curve (Figure 3.5). The longest time delay was observed for an increase in the OLR from a change in the HRT, while the shortest time delay was obtained for a decrease in the COD$_{in}$ concentration (Table 3.2).

The comparison of the PID parameters showed that because of the negligible time delay (t$_d$), especially for the decrease of the COD$_{in}$ concentration, the contribution of the differential and integral part to the PID controller parameters was small, while the proportional gain (K$_c$) parameter was remarkably high, especially for the decrease in OLR as compared to the OLR increase in both OLR change modes (Table 3.2).

TABLE 3.2 Response time, TDS increment (ΔTDS), Cohen-Coon tuning coefficients and PID parameters obtained from a change in the COD_{in} concentration and HRT.

	Change in COD_{in}		Change in HRT	
OLR step change (gCOD/L.d)	0.5-1	1-0.5	0.5-1	1-0.5
Response time (d)	4	2	15	2
ΔTDS	206.5	-180.3	29.8	46.2
K (L.d /g)	-2.2	-0.7	-2.2	1.3
τ (d)	30.8	-51.8	132.4	879.9
t_d (d)	0.3	0.0	2.0	0.4
K_c	-63.7	2538.4	-40.6	3309.6
τi	0.7	0.1	4.9	1.0
τD	0.1	0.0	0.7	0.1

3.4. DISCUSSION

3.4.1. BIOREACTOR RESPONSE TO THE OLR INCREMENT

This study shows that the manipulation of the COD_{in} concentration (for high ΔTDS) or the HRT (for low ΔTDS) in a sulphate reducing bioreactor creates an informative pS response that characterizes the changes in the sulphide concentration. The increase of the OLR by a step change of the COD_{in} concentration showed a high sulphide load change and almost negligible time delays (Table 3.2). An increment in the OLR via a change in the HRT displayed less visible changes in the sulphide response yielding higher integral and derivative PID parameter values as compared to the ones obtained via a change in the COD_{in} concentration (Table 3.2). This was due to the insufficient sulphate reducing rates (for lactate: 16.7 gCOD/gVSS.d, for acetate: 37.1 gCOD/gVSS.d), which were slower as compared to the faster influent flow and thus, the produced sulphide was diluted.

Despite the adequate sulphide response, several days (response time) were needed to achieve pS steady state values in both tuning strategies (Figure 3.5 and Table 3.2) that can destabilize the bioreactor in case of excessive control actions. This can lead to a COD overload and hence, substrate inhibition (Qatibi et al., 1990) or sulphide toxicity (O'Flaherty and Colleran, 2000; Reis et al., 1992) when applying a change in the COD_{in} concentration or biomass washout (Kaksonen et al., 2004) when applying a change in the HRT. The differences in COD and sulphate removal efficiencies, as well as sulphide and acetate production between both OLR change modes for a positive or negative step change (Table 3.1) showed that different metabolic pathways accounting for sulphate reduction and organic matter oxidation were induced (Dunn et al., 2005) depending on the OLR change strategy applied. The substrate utilization by microbial groups other than SRB such as lactate fermentation to acetate (Oyekola et al., 2012) can be favored with the changes in OLR and can also increase the response time even though acetate utilization by SRB occurred, since every degradation step determines the final sulphide concentration at steady state. The decrease of the HRT caused acetate accumulation in the bioreactor (Table 3.1), while a longer HRT allowed acetate utilization (Table 3.1). This is in agreement with Zhang and

Noike (1994) who demonstrated that the HRT is a significant factor in the selection of the predominant microbial species, where acetogens were found to be particularly sensitive to changes in HRT.

3.4.2. BIOREACTOR RESPONSE TO THE DECREASE OF THE OLR

Contrary to the important sulphide increment when the OLR change was positive (Figure 3.5a), decreasing the OLR displayed a less visible change in the pS values (Tuning I- Figure 3.5a) or even a continued sulphide increase (Tuning II- Figure 3.5b), and thus K_c values that could make the bioreactor unstable (Table 3.2).

Longer HRTs (OLR decrease) caused an increase of the sulphide concentration (Figure 3.5b and Table 3.1) as well as an increment in the COD and sulphate removal efficiencies (Table 3.1). Indeed, longer residence times allow for more SRB bioconversion activity when the COD and sulphate are present in excess and thus the limiting factor is the HRT, which in turn allows accumulation of sulphide in the bioreactor.

For the change in the COD_{in} concentration, the small decrease of the sulphide concentration suggests that the sulphide production could be maintained despite the decrease in the COD_{in} concentration. This was not related to an increase in the sulphate reduction rate (Table 3.1) nor a more efficient use of the substrate available because the COD removal efficiency prior to and after the step response for a positive and negative OLR step change was similar (Table 3.1).

It was also not due to a potential difference between the theoretical and real HRT (Figure 3.4) that could delay the sulphide response to the change in OLR. Therefore, it is hypothesized that storage products accumulated during the COD_{in} increase phase that were subsequently consumed when the COD_{in} concentration was decreased. Indeed, SRB (Hai et al., 2004) among other microorganisms (Salehizadeh and van Loosdrecht, 2004) have the capacity to accumulate storage products under feast-famine conditions that can be consumed later by the SRB. The step responses induce periods of excess of carbon alternated with substrate limitation, most probably favoring the selection of biomass with substrate storage capacity (Serafim et al., 2008). This phenomenon has to be considered in the process control of SRB bioreactors as it can affect the control of the sulphide concentration due to non-accounted stored substrate sources.

3.4.3. PRACTICAL IMPLICATIONS TOWARDS THE APPLICATION OF A STRATEGY FOR SULPHIDE CONTROL IN SRB BIOREACTORS

This study provides a baseline of the application of a strategy to control the sulphide concentration in sulphate reducing bioreactors by showing the main drawbacks found to further adapt a PID controller, as adopted in several studies through the modification of conventional controllers for particular systems (Heinzle et al., 1993). The long response time of the system to the step change (Figure 3.5a and Figure 3.5b) is clearly serious drawback. Since acetate is an important parameter that reflects response time and degradation pathways, the output measurement of the acetate concentration is an option to adapt the PID controller for its application in bioreactors to control the sulphide concentration. The volatile fatty acids concentration (including acetate) was incorporated as an output parameter in a control gain configuration of a traditional PI controller to enhance the robustness of the control scheme with respect to influent disturbances in an anaerobic bioreactor regulating the effluent COD (Alvarez-Ramirez et al., 2002).

Another adaptation of the PID parameters can be obtained from the information of the dynamics of the bioprocess such as reaction pathways and kinetics as well as mass balances, which are always required in biological systems due to their non linearity and non stationary characteristics (Steyer et al., 2000). This information can help to predict the response time of the system to the applied change thus preventing an excessive control action. Finally, similarly to the establishment of a maximum permissible pH value in anaerobic processes to avoid inhibition due to volatile fatty acids (Ryhiner et al., 1993), maximum permissible COD_{in} loads should be considered in the controller to avoid biomass inhibition due to excessive sulphide or substrate concentrations.

Another challenge in the application of a control strategy of the pS response refers to the need of an adequate control action if one wants to decrease the sulphide concentration within an acceptable time frame (Figure 3.5a and Figure 3.5b). Therefore, another strategy should be considered over merely decreasing the OLR. One option could be the dilution of the effluent up to the required level for metal precipitation. This strategy is widely used as control action in wastewater treatment plants for obtaining the effluent quality criteria (Metcalf and Eddy, 2002). Nevertheless, for this electron donor-deficient type of wastewaters (Papirio et al., 2013), this strategy would imply substrate losses, which contradicts the aim to efficiently produce solely the stoichiometric sulphide concentrations required for metal sulphide precipitation.

3.4.4. VALIDATION OF THE PS ELECTRODE RESPONSE

This study shows that the online measurement of the sulphide concentration through the pS electrode response can be used to monitor the sulphide concentration online in SRB bioreactors. The use of a pS electrode in bioreactors gives an advantage over off-line methods for sulphide determination, since the sulphide in the system is measured inside the bioreactor avoiding volatilization or oxidation (Hu et al., 2010). For instance, sulphide diffusing out of the zone where it is produced is in part biologically and chemically oxidized to thiosulphate (Brüser et al., 2000; Middelburg, 2000), which could be partly the reason of the variations in the TDS values measured (Table 3.1), in contrast with the fairly stable pS reading values (Figure 3.3).

The pS electrode response is highly sensitive to pH variations (Figure 3.3b), as it influences the predicted values of the sulfur species (Al-Tarazi et al., 2004, Beneš and Paulenová, 1974). Bisulphide (HS⁻) and hydrogen sulphide (H_2S) coexist at pH values below 7, with H_2S the dominant species. Above pH 7, almost all sulphide is in the form of HS⁻ and only in very basic solutions the sulphide exists primarily as free ion (S^{2-}) (Middelburg, 2000).

The voltage response is also less pronounced at high sulphide concentrations (Figure 3.3a) due to the logarithmic response of the pS values (Equation 3.1) to the changes in the TDS concentration (Figure 3.3b). Therefore, errors in TDS measurements above 100 mg/L are more likely to occur when the S^{2-} species is monitored, as the solution becomes saturated in H_2S and therefore the maximum S^{2-} concentration is not attained (Lewis,

2010). Notwithstanding these characteristics, the pS electrode readings can be used for the sulphide measurement in SRB bioreactors as high sulphide concentrations in these systems are not desirable due to sulphide inhibition of the SRB (Chen et al., 2008; Kalyuzhnyi and Fedorovich, 1998). Moreover, the necessary amount of sulphide to precipitate metals for their recovery from wastewaters is rather low as metal concentrations in mine wastewaters typically range between 10 to 250 mg/L (Papirio et al., 2013). Thus, the pS electrode readings will remain within the sensitivity range limit of the pS electrode.

3.5 CONCLUSIONS

· The study of the sulphide response to the variation of the COD_{in} concentration and the HRT provided valuable information, in terms of time-varying behavior, towards the application of a sulphide control strategy in sulphate reducing reactors for metal recovery.

· Delays in the response time and a high control gain were the most critical factors affecting the design of a sulphide control strategy in bioreactors. These were likely caused by the induction of different metabolic pathways in the anaerobic sludge including the accumulation of storage products.

· The pS electrode response is adequate for applications in sulphate reducing bioreactors. However, pH variations and high sulphide concentrations should be carefully observed for correction of the recorded pS values.

REFERENCES

Al-Tarazi M, Heesink ABM, Azzam MOJ, Yahya SA, Versteeg GF (2004) Crystallization kinetics of ZnS precipitation; an experimental study using the mixed-suspension-mixed-product-removal (MSMPR) method. Cryst Res Technol 39: 675-685.

Alvarez-Ramirez J, Meraz M, Monroy O, Velasco A (2002) Feedback control design for an anaerobic digestion process. J Chem Technol Biot 77: 725-734.

APHA, A.P.H.A. (2005) Standard methods for examination of water and wastewater, Washington D.C.

Bang SW, Clark DS, Keasling JD (2000) Engineering hydrogen sulphide production and cadmium removal by expression of the thiosulphate reductase gene (phsABC) from *Salmonella enterica* Serovar Typhimurium in *Escherichia coli*. Appl Environ Microbiol 66: 3939-3944

Beneš P, Paulenová M (1974) Surface charge and adsorption properties of polyethylene in aqueous solutions of inorganic electrolytes. Colloid Polym Sci 252: 472-477.

Bijmans MFM, Buisman CJN, Meulepas RJW, Lens PNL (2011) Comprehensive Biotechnology (Second Edition). Editor-in-Chief: Murray, M.-Y. (ed), pp. 435-446, Academic Press, Burlington.

Brown K, McGreer E, Taekema B, Cullen J (2011) Determination of Total Free Sulphides in Sediment Porewater and Artefacts Related to the Mobility of Mineral Sulphides. Aquat Geochem 17: 821-839.

Brüser T, Lens PNL, Truper H (2000) Environmental technologies to treat sulfur pollution: principles and engineering. Lens, P.N.L. and Hulshoff, P.L. (eds), pp. 47-85, IWA Publishing, London.

Cord-Ruwisch R (1985) A quick method for the determination of dissolved and precipitated sulphides in cultures of sulphate-reducing bacteria. J Microbiol Meth 4: 33-36.

Chen Y, Cheng JJ, Creamer KS (2008) Inhibition of anaerobic digestion process: A review. Bioresource Technology 99: 4044-4064.

Dan TBB, Frevert T, Cavari B (1985) The sulphide electrode in bacterial studies. Water Res 19: 983-985.

Dunn IJ, Heinzle E, Ingham J, Přenosil JE (2003) Biological Reaction Engineering: Dynamic Modelling Fundamentals with Simulation Examples Wiley-VCH Verlag GmbH &Co. KgaA, Weinheim.

Dunn IJ, Heinzle E, Ingham J, Přenosil JE (2005) Biological Reaction Engineering, pp. 161-179, Wiley-VCH Verlag GmbH & Co. KGaA.

Gibert O, de Pablo J, Luis Cortina J, Ayora C (2004) Chemical characterisation of natural organic substrates for biological mitigation of acid mine drainage. Water Res 38: 4186-4196.

Grootscholten T, Keesman K, Lens P (2008) Modelling and on-line estimation of zinc sulphide precipitation in a continuously stirred tank reactor. Sep Purif Technol 63: 654-660.

Gründig B, Krabisch C (1989) Ion-selective microelectrodes. Principles, design and applications. Berlin, Heidelberg, New York, Tokyo: Springer-Verlag, 1986. 346 pp., 153 fig., 42 tab., DM 158,-, ISBN 3-540-16222-4. Acta Biotechnol 9: 210-210.

Gupta A, Flora JRV, Sayles, GD, Suidan MT (1994) Methanogenesis and sulphate reduction in chemostats—II. Model development and verification. Water Res 28: 795-803.

Hai T, Lange D, Rabus R, Steinbüchel A (2004) Polyhydroxyalkanoate (PHA) accumulation in sulphate-reducing bacteria and identification of a class III PHA synthase (PhaEC) in *Desulfococcus multivorans*. Appl Environ Microbiol 70: 4440-4448.

Heinzle E, Dunn I, Ryhiner G (1993), pp. 79-114, Springer Berlin / Heidelberg.

Hu P, Jacobsen LT, Horton JG, Lewis RS (2010) Sulphide assessment in bioreactors with gas replacement. Biochem Eng J 49: 429-434.

Jagadeesh CAP, Sudhaker RDS (2010) Modeling, simulation and control of bioreactors process parameters - remote experimentation approach. Int J Comput Appl 1: 81-88.

Kaksonen AH, Franzmann PD, Puhakka JA (2004) Effects of hydraulic retention time and sulphide toxicity on ethanol and acetate oxidation in sulphate-reducing metal-precipitating fluidized-bed reactor. Biotechnol Bioeng 86: 332-343.

Kaksonen AH, Puhakka JA (2007) Sulphate reduction based bioprocesses for the treatment of acid mine drainage and the recovery of metals. Eng Life Sci 7: 541-564.

Kalyuzhnyi SV, Fedorovich VV (1998) Mathematical modelling of competition between sulphate reduction and methanogenesis in anaerobic reactors. Bioresource Technol 65: 227-242.

König J, Keesman KJ, Veeken A, Lens PNL (2006) Dynamic Modelling and Process Control of ZnS Precipitation. Separ Sci Technol 41: 1025 - 1042.

Lewis AE (2010) Review of metal sulphide precipitation. Hydrometallurgy 104: 222-234.

Marsili-Libelli S, Beni S (1996) Shock load modelling in the anaerobic digestion process. Ecol Model 84: 215-232.

Metcalf, Eddy (2002) Wastewater Enginering: Treatment & Reuse, McGraw-Hill Education (India) Pvt Limited.

Middelburg JJ (2000) Environmental Technologies to Treat Sulfur Pollution - Principles and Engineering. Lens, P.N.L. and Hulshoff Pol, L.W. (eds), pp. 33-46, London : IWA Publishing, 2000. - ISBN 1900222094.

O'Flaherty V, Colleran E (2000) Environmental Technologies to Treat Sulfur Pollution - Principles and Engineering. Lens, P. and Hulshoff Pol, L. (eds), pp. 467-489, IWA Press, London, UK.

Olivet D, Valls J, Gordillo MÀ, Freixó À, Sánchez A (2005) Application of residence time distribution technique to the study of the hydrodynamic behaviour of a full-scale wastewater treatment plant plug-flow bioreactor. J Chem Technol Biot 80: 425-432.

Oyekola OO, Harrison STL, van Hille RP (2012) Effect of culture conditions on the competitive interaction between lactate oxidizers and fermenters in a biological sulphate reduction system. Bioresource Technol 104: 616-621.

Papirio S, Villa-Gomez DK, Esposito G, Lens PNL, Pirozzi F (2013) Acid mine drainage treatment in fluidized-bed bioreactors by sulphate-reducing bacteria: a critical review. Crit Rev Env Sci Tec. 43: 2545-2580

Pind PF, Angelidaki I, Ahring BK, Stamatelatou K (2003) Monitoring and Control of Anaerobic Reactors. Vol. 82.

Qatibi AI, Bories A, Garcia JL (1990) Effects of sulphate on lactate and C2-, C3- volatile fatty acid anaerobic degradation by a mixed microbial culture. Anton Van Lee 58: 241-249.

Reis MAM, Almeida JS, Lemos PC, Carrondo MJT (1992) Effect of hydrogen sulphide on growth of sulphate reducing bacteria. Biotechnol Bioeng 40: 593-600.

Rodrigo MA, Seco A, Ferrer J, Penya-roja JM, Valverde JL (1999) Nonlinear control of an activated sludge aeration process: use of fuzzy techniques for tuning PID controllers. ISA T 38: 231-241.

Ryhiner GB, Heinzle E, Dunn IJ (1993) Modeling and simulation of anaerobic wastewater treatment and its application to control design: Case whey. Biotechnol Progr 9: 332-343.

Salehizadeh H, van Loosdrecht MCM (2004) Production of polyhydroxyalkanoates by mixed culture: recent trends and biotechnological importance. Biotechnol Adv 22: 261-279.

Sampaio RMM, Timmers RA, Xu Y, Keesman KJ, Lens PNL (2009) Selective precipitation of Cu from Zn in a pS controlled continuously stirred tank reactor. J Hazard Mater 165: 256-265.

Serafim L, Lemos P, Albuquerque ME, Reis MM (2008) Strategies for PHA production by mixed cultures and renewable waste materials. Appl Microbiol Biot 81: 615-628.

Steyer JP, Buffière P, Rolland D, Moletta R (1999) Advanced control of anaerobic digestion processes through disturbances monitoring. Water Res 33: 2059-2068.

Steyer JP, Bernet N, Lens PNL, Moletta R (2000) In: Environmental Technologies to Treat Sulfur Pollution / P.N.L. Lens and L.W. Hulshoff Pol. - London : IWA Publishing, 2000. - ISBN 1900222094, pp. 207-235.

Torner-Morales FJ, Buitrón G (2010) The redox potential as the limiting factor to carry out a combined sulphate-reducing / sulphide-oxidizing process in a single SBR In Proceedings of the 12th World Congress on Anaerobic Digestion, October 31 - November 4, Guadalajara, Mexico.

Veeken AHM, Akoto L, Hulshoff Pol LW, Weijma J (2003a) Control of the sulphide (S^{2-}) concentration for optimal zinc removal by sulphide precipitation in a continuously stirred tank reactor. Water Res 37: 3709-3717.

Veeken AHM, de Vries S, van der Mark A, Rulkens WH (2003b) Selective precipitation of heavy metals as controlled by a sulphide-selective electrode. Separ Sci Technol 38: 1-19.

Villa-Gomez D, Ababneh H, Papirio S, Rousseau DPL, Lens PNL (2011) Effect of sulphide concentration on the location of the metal precipitates in inversed fluidized bed reactors. J Hazard Mater 192: 200-207.

Villa-Gomez D, Enright AM, Rini EL, Buttice A, Kramer H and Lens P (2014) Effect of hydraulic retention time on metal precipitation in sulphate reducing inverse fluidized bed reactors. J Chem Technol Biot DOI 10.1002/jctb.4296

Warfvinge P (2009) Process Calculations and Reactor Calculations. Department of Chemical Engineering, LTH.

Yamaguchi T, Yamazaki S, Uemura S, Tseng IC, Ohashi A, Harada H (2001) Microbial-ecological significance of sulphide precipitation within anaerobic granular sludge revealed by micro-electrodes study. Water Res 35: 3411-3417.

Zagury GJ, Kulnieks VI, Neculita CM (2006) Characterization and reactivity assessment of organic substrates for sulphate-reducing bacteria in acid mine drainage treatment. Chemosphere 64: 944-954.

Zhang CT, Noike T (1994) Influence of retention time on reactor performance and bacterial trophic populations in anaerobic digestion processes. Water Res 28: 27-36.

4

Bioprocess Control of
Sulphate Reduction in
an Inverse Fluidized
Bed Reactor: Role of
Microbial Accumulation
and Dynamic
Mathematical Modelling

CHAPTER 4

Bioprocess
Control of
Sulphate
Reduction in an
Inverse Fluidized
Bed Reactor:
Role of Microbial
Accumulation and
Dynamic
Mathematical
Modelling

ABSTRACT

This study aimed at evaluating the possible impact of substrate accumulation (Sulphate and PHB) in microorganisms on bioprocess control of an inversed fluidized bed (IFB) bioreactor. To investigate the impact of substrate accumulation, step feed changes were induced to an IFB bioreactor performing biological sulphate reduction (SR). A first step feed change load set the COD and sulphate influent concentration to zero. Surprisingly, sulphide was still being produced after 15 days of operation without electron donor and sulphate supply. This suggests that accumulated and/or sorbed COD and sulphate supported the continued biological sulphide production. Polyhydroxybutyrate (PHB) was indeed found present in the sludge and could support the sulphate reduction. A second step feed change of adding solely COD (and no sulfate) resulted in the production of sulphide, suggesting that sulphate had accumulated in the IFB sludge. A mathematical model that includes microbial growth, storage products and metabolism of lactate oxidizing SRB was developed, calibrated and validated. This model was able to simulate the accumulation of both PHB and sulphate in the IFB bioreactor.

This chapter has been submitted as:
Cassidy J, Frunzo L, Lubberding HJ, Villa-Gomez DK, Esposito G, Keesman KJ, Lens PNL (2015) Role of Microbial Accumulation in Biological Sulphate Reduction using Lactate as Electron Donor in an Inverse Fluidized Bed Reactor.

4.1 INTRODUCTION

For many years, anaerobic reduction of sulphate by sulphate reducing bacteria (SRB) has been successfully applied for the treatment of sulphate contaminated wastewater from industries on a larger scale. It offers the possibility of an efficient treatment with low operation costs, using various organic and easily utilizable carbon sources (Liamleam and Annachhatre, 2007). The end product is hydrogen sulphide. Hence, this technique is also suitable for treatment of wastewater that contains dissolved metals as well, as the metals can react with the hydrogen sulphide to form metal sulphide precipitates. Under ambient conditions the metals can be subsequently precipitated with the produced hydrogen sulphide and removed as stable precipitates of sulphide.

Controlling the production of sulphide in a sulphate reducing bioreactor is highly relevant to avoid overproduction of H_2S with all its negative economical and environmental effects, especially H_2S toxicity to the anaerobic bacteria and the H_2S concentration that affects the supersaturation and thus, the size of the metal sulphide precipitates (Villa-Gomez et al., 2014). Proportional-integral-derivative (PID) controllers have been successfully used in anaerobic bioreactors. For the design of a PID control strategy that uses the organic loading rate (OLR) as control input, a step response needs to be applied to the bioreactor to obtain the controller parameters from the dynamics of the process (Cassidy et al., 2015 and references therein). For the bacteria, these changes in OLR create feast/famine conditions which induce different metabolic responses as compared to continuous feeding. For SRB, such conditions induce the accumulation of storage compounds, both sulphate and PHB (Cypionka, 1989; Hai et al., 2004). The induction of different metabolic pathways in the anaerobic sludge, including the accumulation of storage products, can result in response delays. Such delays in the response time and a high control gain are the most critical factors affecting the application of a sulphide control strategy (Villa-Gomez et al., 2014).

The aim of this study was to evaluate the possible impact of substrate accumulation (sulphate and PHB) in microorganisms on the design of a control strategy for a continuous sulphate reducing bioreactor. For this purpose, organic shock loads were applied to the bioreactor and the biological response was monitored to determine to what extent the accumulation products could affect the sulphide production as a function of time. In addition, activity batch tests were performed to determine the impact of sulphate accumulation and the feasibility of chemical PHB as electron donor for the species present in the IFB bioreactor. To further understand and test the biochemical pathways proposed in this work and to calibrate crucial parameters affecting the accumulation, a mathematical model describing the processes prevailing in a sulphate reducing bioreactor with microbial accumulation of sulphate and PHB was developed, calibrated and validated.

4.2 MATERIALS AND METHODS

4.2.1 IFB BIOREACTOR AND SHOCK LOADS

Sulphate reduction, using a pS electrode for sulphide control, was performed in an inverse fluidized bed (IFB) bioreactor as described by Villa-Gomez et al. (2014) (Figure 4.1), with the same synthetic medium. Lactate was used as the electron donor and carbon source, sulphate was added as Na_2SO_4 at a COD/SO_4^{2-} ratio of 0.67. The hydraulic retention time (HRT) was kept constant at 1 day.

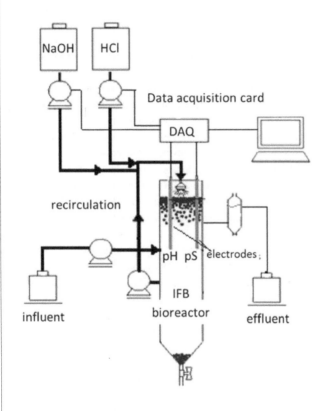

FIGURE 4.1 Experimental set-up with pS and pH control strategy (Villa-Gomez et al., 2014).

Step changes in the OLR were applied to create a response in the sulphide production. To determine the amount of substrate that had been accumulated, the OLR was changed from 1 to 0 g COD/L, using tap water, at time zero. COD, volatile fatty acids, sulphate, sulphide and PHB were monitored daily. Before this step change, the IFB bioreactor had been in operation for 2 years with varying conditions as described in Villa-Gomez et al. (2014). Once the sulphide concentration reached zero, a dynamic COD loading was induced without sulphate addition. In a first phase, a 0.3 gCOD/L load for 1 day was given and when after a couple of days the sulphide had reached zero, the COD load was increased again to 0.3 gCOD/L for the remaining 9 days.

4.2.2 BATCH TESTS

Two groups of batch tests were performed: 1) to determine whether PHB can be used as an electron donor for sulphate reduction and 2) to determine the accumulation of sulphate in the biomass.

4.2.2.1 SULPHATE ACCUMULATION

Activity tests were performed to evaluate sulphate accumulation in the biomass. For this purpose, 117 mL serum bottles were used and 0.2 g/L sulphate was added as Na_2SO_4 at a COD/SO_4^{2-} ratio of 0.67 using lactate as the substrate. The pH was adjusted to 7.0 ($\pm$ 0.2) with NaOH. Each bottle contained 5 mL of carrier material recently taken from the reactor or 2.5 mL of anaerobic granular sludge (Industriewater Eerbeek, Eerbeek, the Netherlands, sampled January 2013) and 15 mL of mineral medium (same composition as the one used for the IFB bioreactor). Both sources of biomass were subjected to a 20 days starvation period previous to the start of the accumulation test to ensure consumption of previously accumulated substrate. The serum bottles were closed with butyl rubber stoppers and flushed with N_2 gas for 3 minutes to achieve anaerobic conditions and to remove any remaining H_2S. The incubations were carried out in triplicate (1) lactate; 2) lactate and sulphate; 3) lactate, sulphate and p-trifluoromethoxy carbonyl cyanide phenyl hydrazone(FCCP) as an uncoupler agent to inhibit sulphate accumulation (Warthmann & Cypionka, 1990) and maintained at 30°C ($\pm$ 2°C) and at a constant rotation of 125 rpm for 21 days. Initial and final sulphate and sulphide concentrations were measured.

To determine the amount of sulphate accumulated, biomass samples were washed with phosphate buffer by centrifugation at 19000 rpm at 4 °C for 20 minutes to remove any remaining sulphate attached to the surface of the biomass. The supernatant was then removed, 5 mL of buffer solution was added and the samples were sonicated at high frequency for 15 minutes to lyse the cells and release the sulphate present, keeping the conditions anaerobic. The sulphate concentration was then measured in the supernatant.

4.2.2.2 PHB AS ELECTRON DONOR

Another set of activity tests was performed to evaluate the feasibility of PHB as an electron donor for sulphate reduction. For this purpose, 117 mL serum bottles were used using PHB and/or lactate as the substrates always maintaining 0.5 gCOD/L and sulphate was added as Na_2SO_4 at a COD/SO_4^{2-} ratio of 0.67. The pH was adjusted to 7.0 ($\pm$ 0.2) with sodium hydroxide. Each bottle contained 5 mL of carrier material recently taken from the reactor or 2.5 mL of anaerobic granular sludge (Industriewater Eerbeek, Eerbeek, the Netherlands, January 2013) and 15mL of mineral medium (same composition as the one used for the IFB bioreactor). The serum bottles were closed with butyl rubber stoppers and flushed with N_2 gas for 3 minutes to achieve anaerobic conditions and to remove any remaining sulphide. The incubations (PHB, PHB+lactate, lactate, no COD) were done in triplicate and maintained at 30°C ($\pm$ 2°C) and at constant rotation of 125 rpm for 21 days. Sulphate, sulphide, PHB, lactate and methane concentrations were determined throughout the experiment.

4.2.3 ANALYTICAL MEASUREMENTS

COD was measured by the closed reflux method (APHA 2005). Sulphate was measured with an ion chromatograph (ICS-1000 Dionex with AS-DV sampler) with a column (IonPac AS14n) at a flow rate of 0.5mL/min with an 8 mM Na_2CO_3/1 mM $NaHCO_3$ eluent, a temperature of 35 °C, a current of 35 mA, an injection volume of 10 μL and a retention time of 8 min. Volatile fatty acids were measured by gas chromatography (GC-CP 9001 Chrompack) after acidification of the samples with 5% concentrated formic acid and filtration through a 0.45 μm nitrocellulose filter (Millipore). The gas chromatograph was fitted with a WCOT fused silica column, the injection and detector temperatures were 175 and 300°C respectively. The temperature of the oven was kept at 115°C. The carrier gas was helium at 100 mL/min. Sulphide was determined spectrophotometrically by the Cord-Ruwisch method and with a pS electrode; the latter was used in the IFB bioreactor only. The VSS content of the sludge was determined according to standard methods (APHA 2005).

Methane (CH_4) was measured with a gas chromatograph (Gas Chromatograph CP 3800) fitted with a PORABOND column Q (25m*0.53mm*10μm) and a TCD detector. The oven temperature was 22°C and the injection volume was 500 μL. The carrier gas was helium at 15 psi.

The presence of PHB was firstly assessed by staining a biomass sample collected from the bioreactor before the the OLR was changed from 1 to 0 gCOD.L⁻¹. The biomass samples were stained with 1% Nile Blue A (Sigma-Aldrich) and visualized in a fluorescent microscope (Olympus BX 51) at excitation and emission wavelengths of 362 and 610 nm, respectively. The quantification of PHB was done with a modified method of Oehmen et al. (2005). Briefly biomass samples were taken from the IFB bioreactor and mixed with formaldehyde to inhibit the microbial activity and frozen until analysis was performed. The samples were centrifuged and the supernatant was removed. All samples were dried under vacuum (-40kPa) and $CaCl_2$. The dried sample was added to 2 mL of chloroform and 2 mL of acidified methanol (containing 3% sulphuric acid as well as 100mg/L sodium benzoate, used as internal standard). Twelve standard solutions were composed of 0 to 3 mg of R-3-hydroxybutyric acid (Sigma-Aldrich). The samples and standards were then digested in tightly sealed glass vials for 2h at 100°C, and cooled to room temperature. Distilled water (1mL) was then added and each sample was mixed vigorously. After mixing, 1h settling time was allowed to achieve phase separation. The chloroform (bottom) phase was then transferred to another vial and dried with anhydrous sodium sulphate. Two microliters of the chloroform phase were analysed with a Agilent Technologies gas chromatograph. The chromatograph was operated with a fused silica capillary column (30m x 0.25mm x 0.25 μm film) and helium as a gas carrier (1 mL/min). The GC-MS system incorporated a similar column coupled with a mass spectrometer 5973.

4.2.4 MODEL DEVELOPMENT

4.2.4.1 NUMERIC INTEGRATION

The numerical integration was performed using MATLAB®'s built-in function ode15s, which is a multi-step, variable-order, solver based on the numerical differentiation formulas.

4.2.4.2 MODEL CALIBRATION AND VALIDATION

A sensitivity analysis (SA) was performed to determine the parameters that affected the model output results the most. These parameters were subsequently calibrated from experimental data. The Matlab® subroutine sens_sys (Bartlett & Davis 2005) was used to perform the parameter SA. This subroutine is an extension of the ode15s' function and calculates the sensitivities of the solution with respect to the parameters by using an iterative approximation based on directional derivatives.

The calibration was performed by applying a response surface methodology (Box et al., 1951; Boels et al., 2012) to find the combination of values of the most sensitive parameters, which gave the lowest root mean square error between the simulated results and steady state experimental measurements of SO_4^{2-} in the effluent. The aim of the validation was to verify the agreement between simulated and experimental data for lactate and sulphide using a new set of experimental data. The comparison between simulated results and experimental measurements was done by applying two methods commonly used for model calibration (Janssen & Heuberger, 1995; Frunzo et al., 2012). The methods used to quantify the error estimation were based on the index of agreement (IoA) and the root mean square error (RMSE):

$$IoA = 1 - \frac{\sum_{i=1}^{K} (y_i - y'_i)^2}{\sum_{i=1}^{K} (|y'_i - y_M| + |y_i - y_M|)^2} \qquad (4.1)$$

$$RMSE = \sqrt{\frac{\sum_{i=1}^{K} (y_i - y'_i)^2}{K}} \qquad (4.2)$$

where y_i is the single simulated value, y'_i is the corresponding observed value, y_M is the average of the simulated values and K is the number of parameters.

4.3 RESULTS

4.3.1 ASSESSMENT OF THE PRESENCE OF PHB IN THE BIOREACTOR

A first assessment was made to verify the presence of PHB in the bioreactor system. This was done by staining a biomass sample collected from the bioreactor before the organic shock loads were applied. Without sonication (Figure 4.2a), PHB was observed in the form of an aggregate. To verify if the PHB was present in association with the cells of the microorganisms or suspended in the liquid, sonication at low frequency was performed in order to destroy the microbial clusters. The PHB granules were associated to the microbial cells (Figure 4.2b) and appeared present in different microbial species, given the different shapes of the cell agglomerates in which PHB was present (Figure 1b). This test gave a clear, albeit only qualitatively, indication that PHB was present in the IFB sulphate reducing sludge.

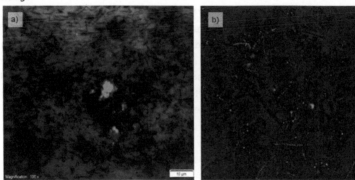

FIGURE 4.2 Nile Blue A stained biomass from the IFB reactor containing PHB before (a) and after sonication at low frequency (20 kHz) (b).

4.3.2 IFB BIOREACTOR RESPONSE TO STEP FEED CHANGES

4.3.2.1 IFB BIOREACTOR RESPONSE TO OMISSION OF COD OR SULPHATE SUPPLEMENTATION

Figure 4.3a shows the evolution of sulphide, sulphate and COD concentration in the effluent of the IFB bioreactor during a step change induced in the organic loading rate at time zero. The OLR was changed from 1 gCOD/L to 0 gCOD/L in order to determine the effect of the substrate accumulation. The response of the bioreactor can be divided in four distinct phases (Figure 4.3a). Phase 1 corresponds to a rapid decrease of COD and sulphate concentrations which can be related to the dilution of the system. Phase 2 shows a steady state where only a small amount of sulphide was produced probably due to an adaptation period of the microorganisms to the organic shock load. In phase 3 the sulphide concentrations increased to values of around 200 mg/L on day 15 and then decreased to values close to zero on day 17. This phase corresponds to the period where all carbon accumulated (as PHB) is consumed (Figure 4.3b).). In the period preceding this step feed change, the only volatile fatty acid present was acetate. After the step feed change, in addition to acetate, butyrate and propionate were also detected (data not shown), which are all degradation products of PHB and other endogenous carbon sources. Although there was still sulphate present in phase 4 (Figure 4.3a), no sulphide was produced. The COD measured was likely the inert fraction and not available for the microorganisms and thus, no sulphate could be reduced by the SRB. Sixteen days were needed for all accumulated carbon (PHB and maybe other forms of stored carbon) to be consumed and reduce the sulphide production to values close to zero. This clearly shows the effect of the accumulation of both carbon and sulphate in the microorganisms.

4.3.2.2 IFB BIOREACTOR RESPONSE TO DYNAMIC COD LOADING

To assess if the lack of non-inert COD was the limiting factor for no sulphide production in phase 4, (Figure 4.3a) COD was added to the system at different time instants. In the first stage, 0.3 gCOD/L during 1 day was added. As soon as COD was added, there was a production of sulphide (Figure 4.3c). Thus, lack of COD was indeed limiting the sulphate reduction, i.e., all accumulated and available carbon sources were degraded in the previous experiment (as seen in Figure 4.3). Since there was still sulphate present and no sulphide was produced, in a second stage 0.3 gCOD/L was added continuously. Similarly, sulphide was produced immediately after the addition of COD, demonstrating again that the lack of carbon source was limiting the biological sulphate reduction.

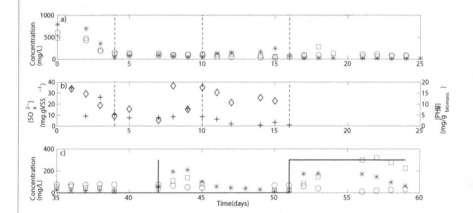

FIGURE 4.3 a)Sulphide (*), COD (□) and sulphate (o) concentration profiles in the effluent of the IFB bioreactor after a step change from 1 gCOD/L to 0 gCOD/L at time zero. Dashed lines divide the different response phases. b) PHB (+) and sulphate released from the cells (◊) concentration present in IFB biomass after a step change from 1 gCOD/L to 0 gCOD/L. Dashed lines divide the different response phases. c) Sulphide (*), COD (□) and sulphate (o) profiles in the effluent during dynamic COD loading (solid line).

4.3.3 MODEL CONSTRUCTION

4.3.3.1 BIOCHEMICAL REACTIONS

The proposed mathematical model takes into consideration the bacterial groups and pathways as shown in Figure 4.4. The mathematical model is described in detail in Table 4.2 and Table 4.3. The model takes two groups of SRB into account classified as incomplete oxidizers (SRBi) and complete oxidizers (SRBc) and eight components (substrates, accumulated compounds and products): Sulphate (SO_4^{2-}), Lactate, Sulphate accumulated ($SO_4^{2-}acc$), PHB, sulphide (HS^-), carbon dioxide (CO_2), acetate and inert (COD generated by the biomass decay). All the components are expressed as COD with the exception of. To correct this, a factor of 1.5 is used in the first column of Table 4.2 representing 1.5g SO_4^{2-}.g COD^{-1}. The kinetic and stoichiometric parameters adopted are listed in Table 4.1.

The model is based on the following assumptions and considerations: (1) the growth of biomass proceeds according to Monod kinetics, (2) the biological reactor is a completely stirred tank reactor (CSTR) with biomass attachment and (3) sulphide inhibition, pH effect, and COD production from biomass decay are not considered.

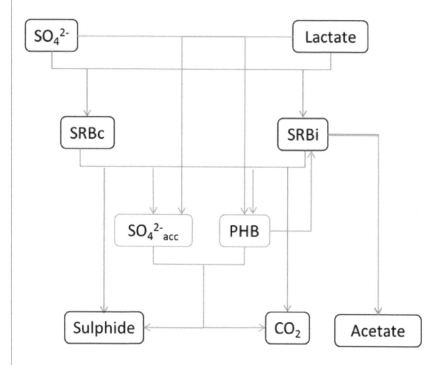

FIGURE 4.4 Schematic representation of the biochemical pathways included in the model. Blue lines correspond to the biochemical pathways without substrate accumulation and orange lines correspond to the biochemical pathways with substrate accumulation.

TABLE 4.1 Kinetic and stoichiometric parameters.

Symbol (units)	Definition	Value	References
μ_{SRBi}^{max} (day^{-1})	Maximum specific growth rate of SRBi	0.51	Kalyuzhnyi and Fedorovich, 1998
K_{Lac}^{SRBi} (gCOD.L^{-1})	Half-saturation coefficient of SRBi on lactate	0.10	Fedorovich et al., 2003
$K_{SO_4^{2-}}^{SRBi}$ (g.L^{-1})	Half-saturation coefficient of SRBi on sulphate	0.0192	Kalyuzhnyi et al., 1998
K_{PHB}^{SRB} (gCOD.L^{-1})	Half-saturation coefficient of SRB on PHB	0.10	Fedorovich et al., 2003
μ_{SRBc}^{max} (day^{-1})	Maximum specific growth rate of SRBc	0.51	Kalyuzhnyi and Fedorovich, 1998
K_{Lac}^{SRBc} (gCOD.L^{-1})	Half-saturation coefficient of SRBc on lactate	0.10	Fedorovich et al., 2003
$K_{SO_4^{2-}}^{SRBc}$ (g.L^{-1})	Half-saturation coefficient of SRBc on sulphate	0.0192	Kalyuzhnyi et al., 1998
S_t (g.L^{-1})	Threshold concentration for accumulated sulphate	0.05	Weijma et al., 2000
S_{tp} (g.L^{-1})	Threshold concentration for PHB	0.003	Ribes et al., 2004
K_d^{SRBi} (day^{-1})	Decay rate of SRBi	0.025	Kalyuzhnyi and Fedorovich, 1998
K_d^{SRBc} (day^{-1})	Decay rate of SRBc	0.025	Kalyuzhnyi and Fedorovich, 1998
Y_{SRB} (gVSS.gCOD^{-1})	Yield of SRB on lactate	0.12	Kalyuzhnyi and Fedorovich, 1998
S_{max} (g.L^{-1})	Maximum apparent sulphate accumulation	0.8	
P_{max} (g.L^{-1})	Maximum apparent carbon accumulation	0.4	
f_d^{max}	Maximum sulphate accumulation factor	0.2	*
f_p^{max}	Maximum PHB accumulation factor	0.9	*

*determined in this chapter

TABLE 4.2 Petersen matrix of the proposed model.

Component → / Process ↓	$S_{SO_4^{2-}}$	$S_{Lactate}$	$S_{SO_4^{2-}\,acc}$	S_{PHB}	S_{HS^-}	$S_{Acetate}$	X_{SRBc}	X_{SRBl}	X_i	Rate ρj (gCOD dm^{-3} day^{-1})
SR Lactate by X_{SRBl}	$-1.5^*(1-Y_{SRB})/(Y_{SRB})$	$-1/Y_{SRB}$	$f_d^*(1-Y_{SRB})/Y_{SRB}$	$f_p^*(1-Y_{SRB})/Y_{SRB}$	$(1-Y_{SRB})/(Y_{SRB})$	$0.8^*(1-f_d)^*(1-Y_{SRB})/(Y_{SRB})$		1		$\mu_1^{inc}\cdot X_{SRBl}$
SR PHB by X_{SRBl}	$-1.5^*(1-Y_{SRB})/(Y_{SRB})$		$f_d^*(1-Y_{SRB})/Y_{SRB}$	$-1/Y_{SRB}$	$(1-Y_{SRB})/(Y_{SRB})$	$0.8^*(1-f_d)^*(1-Y_{SRB})/(Y_{SRB})$		1		$\mu_2^{inc}\cdot X_{SRBl}$
S_{sto}R Lactate by X_{SRBl}		$-1/Y_{SRB}$	$-(1-Y_{SRB})/Y_{SRB}$	$f_p^*(1-Y_{SRB})/Y_{SRB}$	$(1-Y_{SRB})/(Y_{SRB})$	$0.8^*(1-f_d)^*(1-Y_{SRB})/(Y_{SRB})$		1		$\mu_3^{inc}\cdot X_{SRBl}$
S_{sto}R PHB by X_{SRBl}			$-(1-Y_{SRB})/Y_{SRB}$	$-1/Y_{SRB}$	$(1-Y_{SRB})/(Y_{SRB})$	$0.8^*(1-Y_{SRB})/(Y_{SRB})$		1		$\mu_4^{inc}\cdot X_{SRBl}$
SR Lactate by X_{SRBc}	$-1.5^*(1-Y_{SRB})/(Y_{SRB})$	$-1/Y_{SRB}$	$f_d^*(1-Y_{SRB})/Y_{SRB}$	$f_p^*(1-Y_{SRB})/Y_{SRB}$	$(1-Y_{SRB})/(Y_{SRB})$		1			$\mu_1^{com}\cdot X_{SRBc}$
SR PHB by X_{SRBc}	$-1.5^*(1-Y_{SRB})/(Y_{SRB})$		$f_d^*(1-Y_{SRB})/Y_{SRB}$	$-1/Y_{SRB}$	$(1-Y_{SRB})/(Y_{SRB})$		1			$\mu_2^{com}\cdot X_{SRBc}$
S_{sto}R Lactate by X_{SRBc}		$-1/Y_{SRB}$	$-(1-Y_{SRB})/Y_{SRB}$	$f_p^*(1-Y_{SRB})/Y_{SRB}$	$(1-Y_{SRB})/(Y_{SRB})$		1			$\mu_3^{com}\cdot X_{SRBc}$
S_{sto}R PHB by X_{SRBc}			$-(1-Y_{SRB})/Y_{SRB}$	$-1/Y_{SRB}$	$(1-Y_{SRB})/(Y_{SRB})$		1			$\mu_4^{com}\cdot X_{SRBc}$
SO_4^{2-} storage	$-f_d^*(1-Y_{SRB})/(Y_{SRB})$									$\sum_{n=3}^{4}(\mu_n^{inc}\cdot X_{SRBl}+\mu_n^{com}\cdot X_{SRBc})$
PHB storage		$-f_p^*(-1Y_{SRB})/(Y_{SRB})$								$\sum_{n=2,4}(\mu_n^{inc}\cdot X_{SRBl}+\mu_n^{com}\cdot X_{SRBc})$
Decay of X_{SRBl}								-1	1	$K_d^{SRBl}\cdot X_{SRBl}$
Decay of X_{SRBc}							-1		1	$K_d^{SRBc}\cdot X_{SRBc}$

TABLE 4.3 Net specific growth rates reported in Table 4.2

Rate ρj (gCOD dm⁻³ day⁻¹)

μ_1^{inc}	$\mu_{SRBi}^{max} \dfrac{S_{Lac}}{K_{Lac}^{SRBi} + S_{Lac}} \dfrac{S_{SO_4^{2-}}}{K_{SO_4^{2-}} + S_{SO_4^{2-}}}$
μ_2^{inc}	$\mu_{SRBi}^{max} \dfrac{S_{PHB}}{K_{PHB}^{SRB} + S_{PHB}} \dfrac{S_{tp}}{S_{tp} + S_{Lac}} \dfrac{S_{SO_4^{2-}}}{K_{SO_4^{2-}}^{SRBi} + S_{SO_4^{2-}}}$
μ_3^{inc}	$\mu_{SRBi}^{max} \dfrac{S_{Lac}}{K_{PHB}^{SRBi} + S_{Lac}} \dfrac{S_{SO_{4sto}}}{K_{SO_4^{2-}}^{SRBi} + S_{SO_{4sto}}} \dfrac{S_t}{S_t + S_{SO_4^{2-}}}$
μ_4^{inc}	$\mu_{SRBi}^{max} \dfrac{S_{PHB}}{K_{PHB}^{SRB} + S_{PHB}} \dfrac{S_{tp}}{S_{tp} + S_{Lac}} \dfrac{S_{SO_{4sto}}}{K_{SO_4^{2-}}^{SRBi} + S_{SO_{4sto}}} \dfrac{S_t}{S_t + S_{SO_4^{2-}}}$
μ_1^{com}	$\mu_{SRBc}^{max} \dfrac{S_{Lac}}{K_{Lac}^{SRBc} + S_{Lac}} \dfrac{S_{SO_4^{2-}}}{K_{SO_4^{2-}}^{SRBc} + S_{SO_4^{2-}}}$
μ_2^{com}	$\mu_{SRBc}^{max} \dfrac{S_{PHB}}{K_{PHB}^{SRB} + S_{PHB}} \dfrac{S_{tp}}{S_{tp} + S_{Lac}} \dfrac{S_{SO_4^{2-}}}{K_{SO_4^{2-}}^{SRBc} + S_{SO_4^{2-}}}$
μ_3^{com}	$\mu_{SRBc}^{max} \dfrac{S_{Lac}}{K_{PHB}^{SRBc} + S_{Lac}} \dfrac{S_{SO_{4sto}}}{K_{SO_4^{2-}}^{SRBc} + S_{SO_{4sto}}} \dfrac{S_t}{S_t + S_{SO_4^{2-}}}$
μ_4^{com}	$\mu_{SRBc}^{max} \dfrac{S_{PHB}}{K_{PHB}^{SRB} + S_{PHB}} \dfrac{S_{tp}}{S_{tp} + S_{Lac}} \dfrac{S_{SO_{4sto}}}{K_{SO_4^{2-}}^{SRBc} + S_{SO_{4sto}}} \dfrac{S_t}{S_t + S_{SO_4^{2-}}}$
f_d	$f_d^{max} \dfrac{S_{max} - S_{SO_{4sto}}}{K_{SO_4^{2-}}^{SRB} + S_{max} - S_{SO_{4sto}}}$
f_p	$f_p^{max} \dfrac{P_{max} - S_{PHB}}{K_{PHB}^{SRB} + P_{max} - S_{PHB}}$

4.3.3.2 MASS BALANCE EQUATIONS

The following mass balance equations for substrates, products and biomass were considered:

$$\frac{d[S]}{dt} = \frac{q}{V}([S]_{in} - [S]) + \sum_{j=1,...,N} \rho_j v_{j,} [S] (0) = [S]_0 \tag{4.4}$$

$$\frac{d[P]}{dt} = \frac{q}{V}([P]_0 - [P]) + \sum_{j=1,...,N} \rho_j v_{j,} [P](0) = [P]_0 \tag{4.5}$$

$$\frac{d[P_{acc}]}{dt} = \frac{q}{V}([P_{acc}]_0 - \alpha [P_{acc}]) + \sum_{j=1,...,N} \rho_j v_{j,} [P_{acc}](0) = [P_{acc}]_0 \tag{4.6}$$

$$\frac{d[X]}{dt} = \frac{q}{V}([X]_{in} - \alpha [X]) + \sum_{j=1,...,N} \rho_j v_{j,} [X](0) = [X]_0 \tag{4.7}$$

where [S] is the concentration of substrate, $[S]_{in}$ in the concentration of substrate in the influent, q is the flow rate, V_{liq} is the volume of the IFB bioreactor, [P] is the concentration of products, $[P]_0$ is the concentration of products at time zero, $[P_{acc}]$ is the concentration of accumulation products, $[P_{acc}]_0$ is the concentration of accumulation products at time zero, α is the dettachment factor, [X] is the concentration of biomass and $[X]_0$ is the concentration of biomass at time zero.

4.3.3.3 SENSITIVITY ANALYSIS

First estimates for the model parameters were taken from the literature and based on experimental results (Table 4.1). In addition to this, a SA was performed to choose the parameters giving the highest sensitivity to the model results. The SA was performed with respect to all variables (data not shown) but closer attention was given to the three process variables considered for calibration and validation, i.e., SO_4^{2-}, lactate and sulphide.

The parameters f_p^{max} and f_d^{max} were chosen for the calibration as they significantly affected all three process variables (See below).

4.3.3.4 MODEL CALIBRATION

For the calibration the bounds on f_p^{max} and f_d^{max} were set at 0 and 1. This was chosen since values below zero are not physically possible and values greater than 1 would result in negative concentrations, which is again not possible. The SO_4^{2-} concentration in the effluent during the time period between days 766 and 827 was chosen for the calibration (Figure 4a). This time interval was the longest period of steady state observed in the IFB with influent concentrations of 0.8 and 0.5 g.L^{-1} for sulphate and lactate, respectively, and a HRT of 1 day. Figure 4.5a shows the simulated curve that presented the best agreement with experimental data for sulphate in the effluent. The response surface plot (Figure 4.5b) presents the estimated values for these two parameters presenting the minimum value for RMSE (0.0060) between simulation and experimental data. This resulted in an estimate of f_p^{max} of 0.9 and an estimate of f_d^{max} of 0.2 when using values of other kinetic and stoichiometric parameters suggested in the literature (Table 4.1).

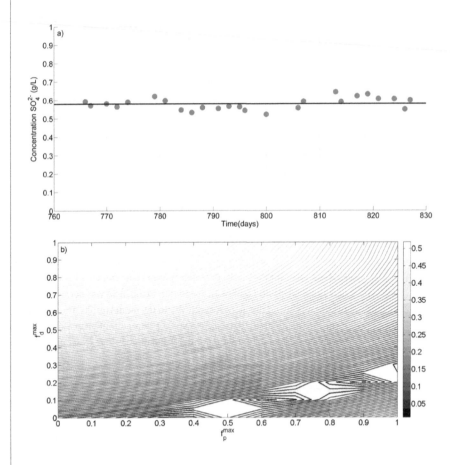

FIGURE 4.5 a) Comparison of simulated and experimental sulphate concentrations. Markers represent experimental data and the full line represents the model simulation. b) Response surface plot presenting the estimated values for f_p^{max} and f_d^{max}. Red mark represents the minimum value for RMSE and corresponding values for f_p^{max} and f_d^{max}.

4.3.3.5 MODEL VALIDATION AND RESULTS

After calibrating the model with the results from sulphate, the model was validated for sulphate, lactate and sulphide in the dynamic time period between days 897 and 918 (corresponding to the dynamic COD loading experiments depicted in Figure 4.3) using the calibrated value of f_p^{max} and f_d^{max}. These validation results (Figures 4.6) show a good agreement between simulated and experimental data for the variables sulphate (RMSE=0.0431; IoA= 0.8106), sulphide (RMSE=0.2100; IoA= 0.9003) and lactate (RMSE= 0.0408; IoA=0.9961).

Figure 4.6d shows the dynamic simulation results for the accumulation of sulphate and PHB in the IFB bioreactor over a period of 1000 days. This simulation was based on the assumption that there was no accumulated sulphate or PHB in the inoculum at time zero. As such, as soon as the reactor started running there was a high increase of storage products due to the spike in external concentrations of sulphate and lactate. The sulphate storage reaches a steady state at around 90 mmol $SO_4^{2-}.gVSS^{-1}$, which is in the same order of magnitude of the values measured in the experimental runs (Figure 4.3). When no external substrate (sulphate and lactate) was added (day 897) both PHB and accumulated sulphate were consumed presenting an inverse profile to the sulphide production (Figure 4.6c). An accumulation of PHB was again observed when the lactate external dosage was kept constant, at 0.3 g.L⁻¹, after day 910.

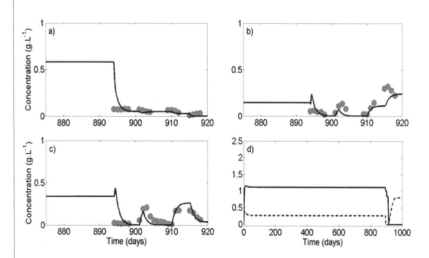

FIGURE 4.6 a) Comparison of simulated and experimental concentrations of a) sulphate, b) lactate (COD) and c) sulphide in the IFB bioreactor effluent. Markers represent experimental data and the full line represents the model simulation. d) Simulation results for sulphate (full line) and PHB (dashed line) accumulation in the IFB bioreactor for a period of 1000 days.

4.3.4 PHB AS AN ELECTRON DONOR - BATCH TESTS

Sulphide was produced in all treatments (Table 4.4) but it was higher in the incubations where a carbon source was present (A, B and C). The effect of PHB and lactate was similar given the similar sulphide production activities (Table 4.4). In some replicas of incubations B and C, the PHB consumption was negative, which suggests there was microbial production of PHB or other storage products during the batch incubation. In incubation D, where no external carbon source was added, there was production of sulphide, which could be caused by stored carbon sources. The IFB biomass showed a higher activity in sulphide production than the anaerobic granular sludge, but no methane production was detected (Table 4.4). On the contrary, the anaerobic granular sludge showed lower sulphide production but there was methane production (Table 4.4). This may be explained by the fact that the IFB biomass was operating under sulphate reducing conditions for 2 years and subjected to several organic step feed changes, whereas the anaerobic granular sludge was used in an experiment shortly after its collection from the Eerbeek wastewater treatment plant.

TABLE 4.4 Sulphide produced, PHB and lactate consumed, and methane produced after addition of A- Sulphate and PHB, B- Sulphate, PHB and lactate, C- Sulphate and lactate, D- Sulphate

		Sulphide produced $(mg.gVSS^{-1}.d^{-1})$	PHB consumed $(mg.mgbiomass^{-1}.d^{-1})$	Lactate consumed $(mgCOD.gVSS^{-1}.d^{-1})$	CH_4 produced $(mg.gVSS^{-1}.d^{-1})$
IFB	A	18.59±3.78	1.04±0.60	n.a.	0
biomass	B	21.77±1.82	0.11±0.59	0.15±0.01	0
	C	16.21±3.59	n.a.	0.20±0.01	0
	D	12.44±1.03	n.a.	n.a.	0
Anaerobic	A	7.66±9.67	0.07±0.01	n.a.	0.63±0.06
granular	B	2.10±0.34	0.04±0.00	0.03±0.01	0.67±0.05
sludge	C	2.65±0.46	n.a.	0.06±0.02	0.97±0.15
	D	1.86±0.26	n.a.	n.a.	0.69±0.06

n.a.-not applicable

4.3.5 SULPHATE ACCUMULATION - BATCH TESTS

Table 4.5 shows the results for the sulphate accumulation batch tests for the IFB biomass and the anaerobic granular sludge. The two sources of biomass show similar trends. However the biomass collected from the IFB bioreactor presents higher accumulation of sulphate, probably due to the fact that it had been subjected to various shock loads during reactor operation, which may have led to a higher accumulation capacity. As expected, there was more accumulation of sulphate in the incubation where sulphate was present and the inhibitor FCCP absent (treatment A). Although the microorganisms were starved previous to the test for 1.5 weeks, a release of sulphate from the cells was also observed in incubation B. This may imply that the starvation period was not sufficient to reduce all sulphate previously accumulated. A very clear difference can be observed between incubations C and the other two incubations. Therefore, the inhibitor for the sulphate binding protein (responsible for the accumulation of sulphate (Warthmann & Cypionka, 1990)) had a great effect on the amount of sulphate accumulated/released from the biomass.

TABLE 4.5 Sulphide produced, sulphate removed and sulphate released from the cells after addition of (A - Lactate and sulphate, B- Lactate, C- Lactate, sulphate and FCCP).

		Sulphide produced $(mg.gVSS^{-1}.d^{-1})$	Sulphate removed $(mg.gVSS^{-1}.d^{-1})$	Sulphate released the cells $(mg.gVSS^{-1}.d^{-1})$
IFB biomass	A	14.08 ±3.52	117.31±9.69	59.98±0.86
	B	20.34 ±3.37	n.a.	48.01±0.06
	C	8.68 ±0.66	73.06±5.53	18.02±0.00
Anaerobic granular sludge	A	0.77 ±0.07	1.75±2.89	4.87±0.00
	B	0.96 ±0.08	n.a.	4.38±0.00
	C	0.53 ±0.00	1.53±0.17	2.10±0.00

n.a.-not applicable

4.4 DISCUSSION

4.4.1 EFFECT OF FEAST-FAMINE CONDITIONS ON MICROBIAL ACCUMULATION

This study showed, to the best of our knowledge, for the first time mixed culture microbial accumulation of both carbon and sulphur sources in continuously operating sulphate reducing bioreactors. This accumulation and consumption of stored compounds will lead to the production of sulphide even when no external substrate is provided. As such, this process is of great importance since it impacts process control strategies for bioreactors with biogenic sulphide production. The results presented here are in agreement with results obtained in a previous study where different tuning strategies were applied for the design of control of the sulphide production (Villa-Gomez et al., 2014).

When the COD and sulphate were removed from the influent, the sulphide production decreased initially, but increased to 0.2 g.L^{-1} and sulphide was still being produced after 15 days of operation without dosage of external substrate (Figure 4.3a). The production of sulphide ceased at the same time as the PHB present in the system was completely consumed (Figure 4.3b)). Sulphate and PHB accumulation was probably a response of the microorganisms to the feast and famine conditions (Hai et al., 2004) induced to this reactor in previous studies (Villa-Gomez et al., 2014). After a period of starvation, the system responded quickly when a lactate load of 0.3 gCOD.L^{-1} for one day was added (Figure 4.3c) leading again to a sulphide concentration of 0.2 g.L^{-1}. Thus, the microbial activity did not seem to be greatly affected by the step feed changes induced in this study. Similarly, when the COD was again dosed in a continuous mode to the system, the sulphide levels returned to 0.2 g.L^{-1}. The accumulation of storage compounds allows the microorganisms to maintain a balanced metabolism under limited substrate conditions and in different feed shocks (van Loosdrecht et al., 1997). The low quantity of PHB present in the cells (Figure 4.3b) confirmed that the microbial groups present were not suitable to develop efficient processes for polyhydroxyalkanoate (PHA, being PHB part of the PHA family) production (Hai et al., 2004). However, given the time needed to reduce sulphide production levels, the occurrence of sulphate and PHB accumulation becomes extremely important to consider when designing a bioprocess control strategy for a sulphate reducing process.

The calibrated and validated model developed in this study showed a high increase of storage in the first days of operation probably due to the spike in external concentration of substrates. The absence of external substrate for a long period of time can cause a decrease in the amount of intracellular components (RNA and enzymes) needed for cell growth of SRB species (Stams et al., 1983). If the microbial culture, after such a starvation period, is spiked with an excess of carbon, the amount of enzymes available in the cells is lower than that required to reach the maximum growth rate (slow growth response). In that case, storage becomes the dominant phenomenon (fast storage response) (Daigger & Grady Jr, 1982).

4.4.2 ACCUMULATION OF SULPHATE

Figures 4.3, 4.5 and 4.6 suggest that the sulphide production is caused by the reduction of internally stored sulphate, thus sulphate appears to be accumulated by the microorganisms present in the IFB biomass in addition to carbon source storage compounds (Figure 4.3, Table 4.5). Sulphate storage is described for other sulphur cycle microbial species such as sulphide oxidizing *Thioploca* (Kojima et al., 2007) and *Beggiatoa* spp., where molar concentrations of sulphate were found in the *Beggiatoa* filaments (Berg et al., 2014). Stored elemental sulphur has been shown to play a key role in the anaerobic oxidation of methane (Milucka et al. 2012) although the S^0 is proposed to accumulate in the ANME cell and not in the SRB cell. There have been also studies which suggest the accumulation of sulphate in SRB in feast and famine conditions as a survival strategy (Cypionka, 1989; Warthmann & Cypionka, 1990): intracellular sulphate concentrations up to 25.6 mM have been reported in *Desulfovibrio desulfuricans* (Cypionka, 1989). This would lead to very high amounts of sulphate accumulated in a running bioreactor. Accumulation of sulphate in cells has been shown in both marine and freshwater SRB species (Cypionka, 1989; Stahlman et al., 1991; Warthmann & Cypionka, 1990).

The activity tests performed give further support to this hypothesis (Table 4.5). In the latter, the accumulation of sulphate was lower in the incubations where an inhibitor of sulphate accumulation (FCCP) was added and higher in the incubations inoculated with the biomass present in the IFB. This suggests that adaptation occurred in the biomass present in the IFB bioreactor after being subjected to different substrate loading conditions, i.e., natural selection of the microorganisms which are able to store sulphate and PHB (Serafim et al., 2008). The accumulation of sulphate was shown to be reversible when higher amounts of sulphate or an uncoupler, dicyclohexylcarbodiimide (DCCD), were added (Cypionka, 1989). However, this was not observed in our study due to the fact that the step change feedings were induced with COD and not with sulphate. In addition, the model developed, taking into account the accumulation, was validated and gave a good fit between simulation and experimental data. The results from the model suggest an intracellular sulphate concentration of 11.54 mM, which is the same order of magnitude as the values measured and presented in Figure 4.3b).

4.4.3 PHB AS AN ELECTRON DONOR FOR SULPHATE REDUCTION

This study showed SRB are able to use PHB as electron donor (Figure 4.3, Table 4.4). Similar activities were observed when using lactate or PHB as carbon source for the microbial growth. PHB was shown to be degraded by pure cultures of SRB (Janssen & Schink, 1993; Çetin, 2009) and by mixed cultures of SRB (Urmeneta et al., 1995). In this study, we showed that PHB can be biodegraded by mixed cultures of SRB and its degradation may lead to the production of intermediates that may be used by methanogens to produce methane (Table 4.4). However, the latter was only observed in the incubations inoculated with anaerobic granular sludge. The consumption of PHB was higher in the experiments inoculated with biomass from the IFB bioreactor. This, together with the results from the sulphate accumulation tests (Table 4.5), gives stronger evidence that the microorganisms in the IFB had adapted to step change feeding and that they accumulated and consumed storage compounds. SRB, however, are not suitable for establishing efficient processes for polyhydroxyalkanoates production due to their slow growth and low cell yields (Hai et al., 2004). This process becomes nevertheless extremely interesting and important when designing a bioprocess control strategy for sulphate reducing bioreactors. The main limiting factor when building a large scale biological treatment process is the electron donor cost. It is attractive to make use of the PHB and other possible storage polymers, such as polyglucose (Stams et al., 1983), that accumulate during such feast/famine conditions as electron donor, and thus reduce the amount of electron donor that needs to be supplied to the reactor. Although it has not been tested for SR, decoupled substrate addition was selected for microorganisms removing nitrogen from wastewater (Scherson et al., 2013). The latter accumulated PHB during an anaerobic period, which was consumed during the subsequent anoxic period for NO_2^- reduction. The biological nitrogen removal efficiency from the water was 98% over more than 200 cycles. Thus, optimization of decoupled substrate addition for SRB processes could be an attractive way of reducing the overall costs of SRB bioreactor systems.

4.5 CONCLUSIONS

· The feast and famine conditions induced the development of a microbial community in the sulphate reducing IFB bioreactor that is capable of storing a carbon and sulphur source as, respectively, PHB and sulphate.

· The understanding of the microbial accumulation of storage products is crucial when determining the amount of electron donor to add to a sulphate reducing bioreactor. This accumulation process may lead to significant decreases in the costs associated to electron donor dosage.

· The metabolic pathways that lead to an accumulation of storage products need to be included in the design of a control strategy for biogenic sulphide production with electron donor dosage as the control input.

· The calibrated and validated model proposed in this study can be used to understand the effect of different operational conditions on the reduction of sulphate, the oxidation of lactate and the accumulation of sulphate and PHB. Not only can this model be used to accurately model dynamic conditions, but it can also be used to optimize the addition of electron donor. Consequently, adequate strategies for the bioprocess control of sulphide production and minimization of electron donor can be developed.

REFERENCES

APHA, A.P.H.A. (2005) Standard methods for examination of water and wastewater, Washington D.C.

Bartlett DW, Davis ME. (2006) Insights into the kinetics of siRNA-mediated gene silencing from live-cell and live-animal bioluminescent imaging. Nucleic Acids Res. 34: 322-333.

Berg JS, Schwedt A, Kreutzmann AC, Kuypers MMM, Milucka J (2014). Polysulfides as intermediates in the oxidation of sulfide to sulfate by *Beggiatoa* spp. App. Environ. Microbiol. 80:629-636.

Boels L, Keesman KJ, Witkamp GJ (2012). Adsorption of phosphonate antiscalant from reverse osmosis membrane concentrate onto granular ferric hydroxide. Environ. Sci. Technol. 46:9638–9645.

Box GEP, Wilson KB. (1951) On the experimental attainment of optimum conditions. J. R. Stat. Soc. B 13:1–45.

Cassidy J, Lubberding HJ, Esposito G, Keesman KJ, Lens PNL. (2015) Automated biological sulphate reduction: a review on mathematical models, monitoring and bioprocess control. FEMS Microbiol Rev. 39(6):823-53

Çetin D (2009) Anaerobic Biodegradation of Poly-3-Hydroxybutyrate (PHB) by Sulfate Reducing Bacterium *Desulfotomaculum* sp. Soil Sediment Contam 18: 345–353.

Cord-Ruwisch, R., (1985). A quick method for the determination of dissolved and precipitated sulfides in cultures of sulfate reducing bacteria. J. Microbiol. Methods 4:33-36.

Cypionka H (1989) Characterization of sulfate transport in *Desulfovibrio desulfuricans*. Arch Microbiol 152:237–243.

Daigger GT, Grady Jr CPL (1982) The dynamics of microbial growth on soluble substrates: A unifying theory. Water Res 16:365–382.

Fedorovich V, Lens P, Kalyuzhnyi S (2003) Extension of Anaerobic Digestion Model No . 1. Appl Biochem Biotechnol 109:33–45.

Frunzo L, Esposito G, Pirozzi F, Lens P (2012) Dynamic mathematical modeling of sulfate reducing gas-lift reactors. Process Biochem 47: 2172–2181.

Hai T, Lange D, Rabus R, Steinbu A (2004) Polyhydroxyalkanoate (PHA) accumulation in sulfate-reducing bacteria and identification of a class III PHA synthase (PhaEC) in *Desulfococcus multivorans*. Appl Environ Microbiol 70: 4440–4448.

Janssen PH, Schink B (1993) Pathway of anaerobic poly-beta-hydroxybutyrate degradation by *Ilyobacter delafieldii*. Biodegradation 4: 179–185.

Janssen PHM, Heuberger PSC (1995) Calibration of process-oriented models. Ecol Modell 83: 55–66.

Kalyuzhnyi S, Fedorovich V, Lens P, Hulshoff Pol L, Lettinga G (1998) Mathematical modelling as a tool to study population dynamics between sulfate reducing and methanogenic bacteria. Biodegradation 9: 187–199.

Kalyuzhnyi S V, Fedorovich VV (1998) Mathematical modelling of competition between sulphate reductions and methanogenesis in anaerobic reactors. Bioresour Technol 65: 227–242.

Kojima H, Nakajima T, Fukui M. (2007). Carbon source utilization and accumulation of respiration-related substances by freshwater *Thioploca* species. FEMS Microbiol Ecol 59:23–31.

Lewis AE (2010) Review of metal sulphide precipitation. Hydrometallurgy 2: 222-234.

Liamleam W, Annachhatre AP (2007) Electron donors for biological sulfate reduction. Biotechnol Adv 25: 452–463.

Milucka J, Ferdelman TG, Polerecky L, Franzke D, Wegener G, Schmid M, Lieberwirth I, Wagner M, Widdel F & Kuypers MMM . 2012. Zero-valent sulphur is a key intermediate in marine methane oxidation. Nature 491: 541–546.

Oehmen A, Keller-Lehmann B, Zeng RJ, Yuan Z, Keller J (2005) Optimisation of poly-beta-hydroxyalkanoate analysis using gas chromatography for enhanced biological phosphorus removal systems. J Chromatogr A 1070: 131–136.

Ribes J, Keesman K, Spanjers H (2004) Modelling anaerobic biomass growth kinetics with a substrate threshold concentration. Water Res 38: 4502–4510.

Scherson YD, Wells GF, Woo S-G, Lee J, Park J, Cantwell BJ, Criddle CS (2013) Nitrogen removal with energy recovery through N_2O decomposition. Energy Environ Sci 6: 241.

Serafim LS, Lemos PC, Albuquerque MGE, Reis M a M (2008) Strategies for PHA production by mixed cultures and renewable waste materials. Appl Microbiol Biotechnol 81: 615–628.

Stahlman J, Warthmann R, Cypionka H. (1991). Na^+-dependent accumulation of sulfate and thiosulfate in marine sulfate-reducing bacteria. Arch Microbiol 155:554-558.

Stams FJM, Veenhuis M, Weenk GH, Hansen TA (1983) Occurrence of polyglucose as a storage polymer in *Desulfovibrio* species and *Desulfobulbus propionicus*. Arch Microbiol 136: 54–59.

Urmeneta J, Mas-Castella J, Guerrero R (1995) Biodegradation of Poly-(beta)-hydroxyalkanoates in a lake sediment sample increases bacterial sulfate reduction. Appl Environ Microbiol 61: 2046.

van Loosdrecht M, Pot M, Heijnen J (1997) Importance of bacterial storage polymers in bioprocesses. Water Sci Technol 35: 41–47.

Villa-Gomez DK, Cassidy J, Keesman KJ, Sampaio R, Lens PNL (2014) Sulfide response analysis for sulfide control using a pS electrode in sulfate reducing bioreactors. Water Res 50: 48–58.

Warthmann R, Cypionka H (1990) Sulfate transport in *Desulfobulbus propionicus* and *Desulfococcus multivorans*. Arch Microbiol 154: 144–149.

Weijma J, Hulshoff Pol LW, Stams AJM, Lettinga G. (2000). Performance of a thermophilic sulfate and sulfite reducing high rate anaerobic reactor fed with methanol. Biodegradation 11:429-439.

5

Effect of Alternative Co-substrates on the Rate of Anaerobic Oxidation of Methane and Sulphate Reduction

CHAPTER 5

Effect of Alternative
Co-substrates
on the Rate of
Anaerobic
Oxidation of
Methane and
Sulphate Reduction

ABSTRACT

Anaerobic oxidation of methane (AOM) coupled to sulphate reduction (SR) is mediated by a consortium of anaerobic methanotrophic archaea (ANME) and sulphate reducing bacteria (SRB). Although this has been studied for many years, the metabolic interactions and pathways involved are still unclear. In this work ANME-2 (ANME group 2)/SRB enriched biomass was incubated under high methane-pressure with several labelled and non-labelled substrates and several labelled and non-labelled substrates and the AOM and SR activities were compared individually for each group based on the production of CO_2 and sulphide. Acetate was the tested compound with the highest positive effect on the SR activity (reducing the lag phase) without a visible effect on the AOM. To understand which pathway is taking place for the acetate oxidation, isotopically labelled carbon compounds were added, i.e. (1) non-labelled acetate (together with $^{13}CH_4$), (2) 2-^{13}C labelled together with $^{12}CH_4$ and (3) $^{13}C_2$ labelled together with $^{13}CH_4$. Group 2 presented the lowest lag phase, possibly due to the preference of the ANME to $^{12}CH_4$ over $^{13}CH_4$. Given the higher amount of $^{13}CO_2$ produced in group 3, it is likely that most $^{13}CO_2$ comes from the oxidation of the non-methyl carbon of the acetate. This paper shows that the ANME-2 enriched community present in the inoculum has diverse pathways, being able to use sulphate, thiosulphate, sulphur as electron acceptor, and methane, acetate as electron donor. In addition, it gave an indication that ANME-2 may oxidize methane anaerobically via an undefined mechanism where SRB is not required as partner.

This chapter will be submitted as:
Cassidy J, Zhang Y, Lubberding HJ, Esposito G, Xiao X, Lens PNL (2015) Effect of Alternative Co-substrates on the Rate of Anaerobic Oxidation of Methane and Sulphate Reduction.

5.1 INTRODUCTION

In marine environments, dissimilatory sulphate reduction (SR) plays a key role in the oxidation of organic matter due to its high concentrations in sea water (Jørgensen and Kasten, 2006). When the oxidants (O_2, NO_3^-, Fe(III), Mn(IV) and SO_4^{2-}) are depleted in the sediment, CO_2 becomes the oxidant of choice, and decomposition of organic matter is linked to CH_4 production (Valentine, 2002). The sediment depth where sulphate reduction gives way to methanogenesis is known as the sulphate to methane transition (SMTZ). Anaerobic oxidation of methane (AOM), which is thought to be responsible for the oxidation of 90% of methane produced in marine sediments (Reeburgh, 2007), occurs in the SMTZ and is believed to be coupled to (SR) according to the following net equation:

$$CH_4 + SO_4^{2-} \rightarrow HCO_3^- + HS^- + H_2O \qquad (5.1)$$

AOM-SR is thought to be mediated by three groups of anaerobic methane-oxidizing archaea (ANME); ANME-1, ANME-2 and ANME-3. ANME are distantly related to the *Methanosarcinales* clade of methanogenic archaea and can be found syntrophically associated with sulphate reducing bacteria (SRB) or exist as single cells (Hinrichs et al., 1999; Boetius et al., 2000; Orphan et al., 2001; Knittel and Boetius, 2009; Tang et al., 2013; Wang et al., 2014). However, there is still a knowledge gap on how these two processes, i.e., oxidation of methane and reduction of sulphate are linked to each other. It is hypothesised that methane is oxidized by ANME to an intermediate which is subsequently used by the SRB to reduce sulphate. On the other hand, by consuming the produced intermediate, SRB maintain it at a very low concentration, keeping AOM energetically favourable (Beal et al., 2011).

The genes required to perform all seven steps of methanogenesis from CO_2 were found present and are actively expressed in ANME-2a (Wang et al., 2014). Thus, it is likely that AOM is carried out through a complete reversal of methanogenesis from CO_2. In the same study, it was shown that ANME-2a own several electron transfer pathways that may allow the microorganism to be more flexible in substrate utilization and they thus, have the capacity for rapid adjustment to changes of the environmental conditions.

Several compounds have been tested to assess their effect on AOM-SR. However divergent results have been achieved. Adding hydrogen, formate, acetate, methanol, carbon monoxide or methylamines reduced sulphate reduction rates in a sediment from Hydrate Ridge, suggesting that the SRB were not adapted to those substrates (Nauhaus et al., 2002, 2005). Similarly, Sørensen et al. (2001) excluded hydrogen, acetate and methanol as intermediates in the AOM-SR process saying that the maximum diffusion distances of the latter compounds, at *in situ* concentrations and rates were smaller than the thickness of two prokaryotic cell walls. Meulepas et al. (2010c) excluded acetate, formate, methanol, carbon monoxide and hydrogen as intermediary compounds as their concentration exceeded 1000x the concentrations at which no more Gibbs free energy can be conserved from their production during methane oxidation at the applied conditions.

Although hydrogen and formate were excluded as they can not be exchanged fast enough between syntrophic partners, as shown by a process-based model, to sustain SR rates

(Nauhaus et al., 2007), it was shown that it can occur for acetate (Orcutt and Meile, 2008). On another study, using a spherical diffusion-reaction model, hydrogen, formate and acetate were found to be thermodynamically and physically possible intermediates in AOM-SR (Alperin and Hoehler, 2010). A recent study has discussed the possibility that AOM might not be an obligate syntrophic process, but may be carried out by the ANME alone with zero-valent sulphur being an intermediate then which is disproportionated by the SRB (Milucka et al., 2012). This disparity in results is partly due to the involvement of different groups of ANME and SRB with different metabolic capacities and/or different environmental conditions. This study presents the effect of several labelled and non-labelled substrates added an ANME-2 enriched AOM-SR inoculum on the AOM and SR activities in high pressure vessels and low temperature (mimicking the original environment). These rates were compared individually for each group based on the CO_2 and sulphide production.

5.2 METHODOLOGY

5.2.1 ORIGIN OF BIOMASS

The biomass used in this study originates from the Gulf of Cadiz in the Atlantic Ocean near the coast of southern Iberian Peninsula. Collection took place in Captain Aryutinov Mud Volcano (coordinates: 35:39.700 / 07:20.012) at 1200m depth on 30th of April 2006 (Zhang et al., 2010). Before its usage for this study, the biomass was enriched in ANME-2/SRB in a continuous high pressure bioreactor at 15°C, simulating a cold seep ecosystem where sulphate and high pressure methane are supplied (Zhang et al., 2010).

5.2.2 BASAL MEDIUM

The basal medium consisted of NaCl 26 g/L, $MgCl_2 \cdot 6H_2O$ 5 g/L, $CaCl_2 \cdot 2H_2O$ 1.4 g/L, NH_4Cl 0.3 g/L, KH_2PO_4 0.1 g/L, KCl 0.5 g/L, Na_2SO_4 1.43 g/L, a bicarbonate solution 30 ml, a trace element solution 1 ml, a vitamin mixture solution 1 ml, a thiamine solution 1 ml, and a vitamin B12 solution 1 ml. The bicarbonate solution, the trace element solution and the vitamin solutions were made according to Widdel and Bak (1998). Resazurine was added to check whether the conditions were anaerobic.

5.2.3 INCUBATION PROCEDURE

To assess the effect of different compounds in the AOM/SR rates, 80 mL high pressure vessels were used to maintain 100 bar pressure (Figure 5.1). In an anaerobic glove box, 16 mL of inoculum and the additional substrate were added to each vessel which were closed and removed from the glove box. The vessels were then flushed with $^{13}CH_4$, $^{12}CH_4$, $^{13}CO_2$ or N_2 and given 3 bar gas pressure. Subsequently, basal medium was added using a high pressure pump (HPLC pump) in order to reach 100 bar pressure. The incubation was performed at 15°C in the dark. The following incubations were performed (1) $^{13}CH_4 + SO_4^{2-}$ (Control), (2) $^{13}CH_4 + SO_4^{2-} + S^0$, (3) $^{13}CH_4 + SO_4^{2-} + S_2O_3^{2-}$, (4) $^{13}CH_4 + SO_4^{2-} + NO_3^-$, (5) $^{13}CH_4 + SO_4^{2-}$ + HCO_2^-, (6) $^{13}CH_4 + SO_4^{2-} + ^{13}C_2H_3O_2^-$, (7) $^{13}CH_4 + SO_4^{2-} + ^{13}CH_3CO_2^-$, (8) $^{12}CH_4 + SO_4^{2-} + ^{12}C_2H_3O_2^-$, (9) $^{12}CH_4 + SO_4^{2-} + ^{13}CO_2$, (10) $^{13}CH_4 + SO_4^{2-}$ + Antibiotics. In addition two different incubations were made, one without inoculum and one without substrate.

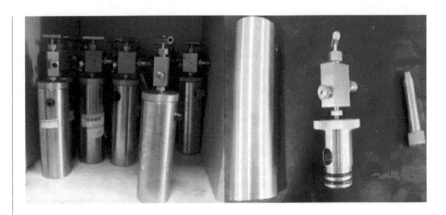

FIGURE 5.1 High pressure metallic vessels used for batch incubations at 100 bar.

All experiments were done in triplicate however, to facilitate the comparison between all treatments only the replica with best behaviour, i.e., higher activity, is shown (all replica results are shown in the Supplementary information).

5.2.4 SAMPLING

For the sampling, each vessel was homogenized. Approximately 1 mL sample was taken by attaching a connector and a vacuum tube to the exit port while gently opening the tap. Weight and pressure were measured in the vacuum tube before and after the sampling. Pressure in each vessel was restored by adding fresh basal medium using the HPLC pump.

5.2.5 CHEMICAL AND BIOLOGICAL ANALYSIS

Dissolved sulphide was measured by using the methylene blue method (Hach Lange method 8131) and a DR5000 Spectrophotometer (Hach Lange GMBH, Düsseldorf, Germany). Sulphate was measured with an ion chromatograph (Metrohm 732 IC Detector) with a column (METROSEP A SUPP 5 - 250) at a flow rate of 0.7mL/min with an 3.2 mM Na_2CO_3 / 1.0 mM $NaHCO_3$ eluent, a temperature of 35 °C, a current of 13.8 mA, an injection volume of 20 µL and a retention time of 35 min.

The headspace composition was measured on a gas chromatograph–mass spectrometer (GC-MS Agilent 7890A-5975C). The system incorporated a DB-5MS column (30 m×0.25 mm× 0.25 µm) MS. Helium was the carrier gas at a flow rate of 1.0 mL min^{-1}. The column temperature was 40 °C.

For acetate measurements, 100 µL of sample were mixed with 300 µL 5% H_2SO_4 in methanol. After 4 hours incubation at 80 °C, 300 µL of 0.9% NaCl and 200 µL of n-hexane were added. The top layer was injected to the Agilent 7890A-5975C GC-MS. The GC-MS column was a DB-5ms (0.25 µm x 0.25 mm x 30 m).

Total ATP was determined using the BacTiter-Glo system (Promega). The luminescence of each sample was measured in a luminometer. Dilutions of pure rATP (Promega) were used to obtain the calibration curve before each experiment.

5.2.6 CALCULATIONS

The dissolved methane and carbon dioxide, and gaseous H_2S were calculated according to the Duan's models, taking pressure, temperature and salinity into account (Duan and Sun, 2003; Duan and Mao, 2006; Duan et al., 2007). Speciation of sulphide (S^{2-}, H_2S, HS^-) and carbonate species (Total Inorganic Carbon - TIC) was determined using the dissociation constants.

5.3 RESULTS

5.3.1 EFFECT OF ALTERNATIVE SUBSTRATES ON SULPHATE REDUCTION

Sulphide production in the high pressure batch tests with different substrates is shown in Figure 5.2. Even though none of the treatments presented higher sulphide production in comparison with the control, some decreased the initial lag phase while some others seem to have inhibited sulphide production. The combinations with $^{13}CO_2$ and NO_3^- ceased sulphide production but it is not clear if it was due to inhibition or microbial competition for the substrate. When antibiotics were added, the sulphide production decreased largely but did not stop completely, which might be because the amount of antibiotics added was insufficient to completely inhibit the microbial activity or that the ANME are capable of reducing small amounts of sulphate themselves. The groups with $S_2O_3^{2-}$, S^0 and HCO_2^- started producing sulphide earlier than the control but reached an earlier steady state with concentrations around 20x, 3x and 2x, respectively, lower than the control. It might be that there were inhibitory compounds formed in the process such as polysulphides or that the thermodynamic conditions were affected. Acetate addition in all forms showed similar end concentrations of sulphide as the control but reduced significantly the lag phase which indicates that the sulphate reducers present in this inoculum are well adapted to acetate. Interestingly, the sulphate removal was one order of magnitude higher than the corresponding sulphide production (Figure 5.3). Thiosulphate was produced in all groups where sulphate was reduced, except for nitrate. The exact pathway for the thiosulfate conversion in this microbial consortium is not yet known. As oxygen was absent (verified by the resazurine), it might be that sulphate is reduced to thiosulphate and perhaps to another unknown sulphur intermediate not analysed as the mass balance is not fully closed. Interestingly, there was a lower production of thiosulphate in the groups where acetate was added even though these groups present higher amount of sulphate reduced (Figure 5.3). This might indicate that the sulphate reducers present in the utilized microbial community do not reduce sulphate completely if the concentration of the electron donor, i.e., acetate, is low.

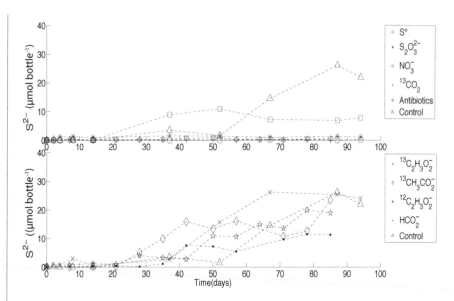

FIGURE 5.2 Sulphide production in time in batch incubations in the absence (control) or presence of co-substrate.

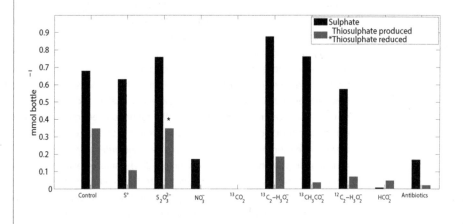

FIGURE 5.3 Sulphate removal, thiosulphate production and thiosulphate removal (*) in batch incubations in the absence (control) or presence of co-substrate.

5.3.2 EFFECT OF ALTERNATIVE SUBSTRATES ON ANAEROBIC OXIDATION OF METHANE

The production of ^{13}C-TIC was used as measurement for the ^{13}CH$_4$ oxidation and is shown in Figure 5.4. None of the substrates added increased the production of ^{13}C-TIC. Contrary to what was observed for sulphide production, there was no difference in lag phase between the control and other groups and ^{13}CH$_4$ oxidation had a sorter lag phase (below 4 days) than the corresponding lag phase for sulphide production (around 30 days) in the control group. The groups with acetate showed no enhancement but also no inhibition in ^{13}C-TIC production. The ^{13}CH$_4$ oxidation seems to have been inhibited by the same compounds inhibiting sulphide production, i.e., antibiotics, nitrate and thiosulphate, probably due to a change in the thermodynamic conditions. The group incubated with ^{13}CH$_3$CO$_2^-$ showed very low production of ^{13}C-TIC, indicating the possibility that the carboxylic group was used by the SRB and methyl group was probably used for methanogenesis. ^{13}C-TIC was consumed in the group where it was added during the first 20 days, it is more likely that it led mainly to the production of methane, as no ^{13}C-acetate was significantly produced in this group (Figure 5.5).

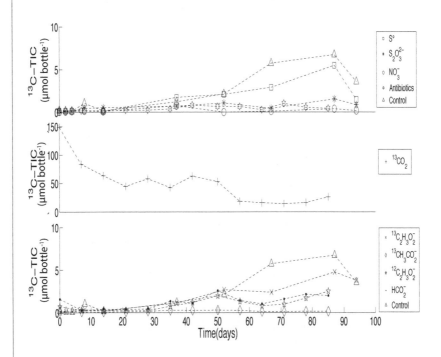

FIGURE 5.4 ^{13}C-TIC production in time in batch incubations in the absence (control) or presence of co-substrate.

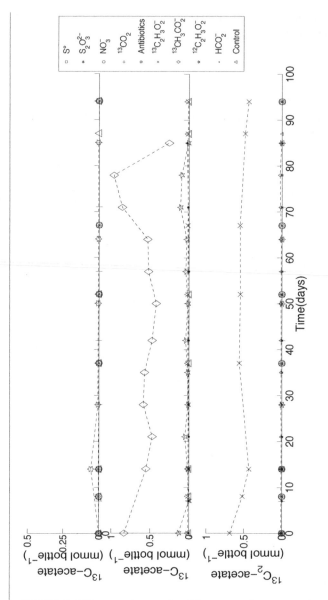

FIGURE 5.5 ^{13}C-acetate (Top two) and ^{13}C$_2$-acetate (bottom) production in time in batch incubations in the absence (control) or presence of co-substrate.

Similarly, there is a consumption of ^{12}C-TIC in all groups. A steep decrease is visible in the first week of incubation which is coinciding with an increase of ^{12}C-acetate (Figure 5.6) which suggests a correlation between these two compounds. Figure 5.6 shows that ^{12}C-acetate decreases in most groups after its production and has a higher production in the group with antibiotics. This suggests that bacterial species are responsible for the oxidation of acetate and thus, probably related to sulphate reduction. In addition, the fact that the ^{13}C-TIC production is lower in the group supplemented with antibiotics suggests a key role of the bacterial partners in maintaining the ideal thermodynamic conditions.

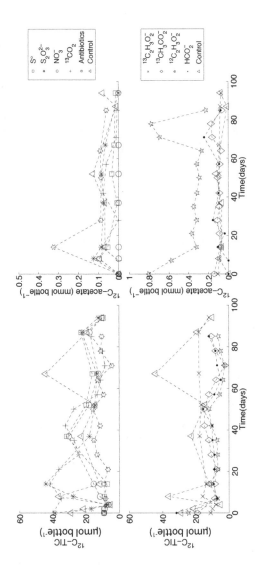

FIGURE 5.6 ^{12}C-TIC and ^{12}C- acetate production in time in batch incubations in the absence (control) or presence of co-substrate.

ATP was measured to indirectly assess the growth or decay of biomass in the different bottles with the different added co-substrates. The ATP was measured at the startup and at the end of the experiments. The results shown in Figure 5.7 corroborate the results presented in sections 5.3.1 and 5.3.2. The group with antibiotics shows the largest decrease in ATP, as expected. There is an increase in ATP in the acetate and control groups. There is a big decrease in ATP in the third replica with sulphur indicating some sort of inhibition. Interestingly, there is an increase in ATP in the group with thiosulphate which produced only small amounts of total inorganic carbon and sulphide, which suggests that other microbial pathways are active with this electron acceptor as well.

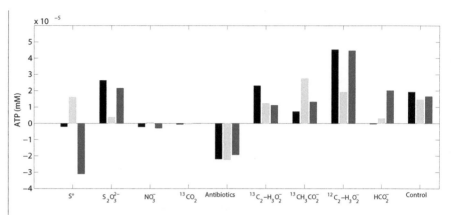

FIGURE 5.7 ATP production (positive) and decrease (negative) for each replica (3 in total) in batch incubations in the absence (control) or presence of co-substrate.

5.4 DISCUSSION

5.4.1 UNCOUPLING OF AOM AND SR

The results obtained clearly showed an uncoupling between sulphate reduction and anaerobic oxidation of methane. The sulphide production was around 3x higher than what would be expected theoretically from the TIC produced. Even more discrepancy is observed when comparing the ^{13}C-TIC results with the amount of sulphate reduced, around 50x more sulphate reduced than what is expected theoretically. In fact, the predicted 1:1 ratio from the AOM-SR reaction overall stoichiometry (equation 5.1) is only rarely observed for AOM and SR rates in AOM-SR sediments (Beal et al., 2011; Bowles et al., 2011). This might be due to other microbial processes taking place.

The results suggest that methanogenesis is also taking place in the ANME-2 enriched AOM-SR inoculum. Many publications suggest that the ANME group perform methanogenesis besides the anaerobic oxidation of methane (Orcutt et al., 2008; Meulepas et al., 2010a; Bowles et al., 2011). Orcutt et al. (2008) showed that the rate of methanogenesis was around 10% of the AOM rate in ANME-2/DSS aggregates. Even though there still hasn´t been confirmation of ANME reversing their metabolic pathways to generate energy from methanogenesis, in a previous work, all the genes required to perform all seven steps of methanogenesis from CO_2 have been found present and actively expressed in the ANME-2a used in this study (Wang et al., 2014). Thus, it could be possible that AOM is carried out through a complete reversal of methanogenesis from CO_2.

5.4.2 ENHANCEMENT OF AOM COUPLED TO SR USING OTHER SUBSTRATES

Although the inoculum was not fed previously with other substrates than sulphate and methane it was able to utilize several other substrates, such as sulphur, thiosulphate and acetate. The use of other co-substrates might enhance the growth rate of these anaerobic consortia as anaerobic oxidation of methane yields a very low Gibbs energy (Reeburgh, 2007). Acetate was clearly the most favourable addition as it reduced the initial lag phase of SR without affecting the AOM. In addition, the inoculum was able to utilize formate, thiosulphate and sulphur. However, the incubations with these three substrates, showed signs of inhibition after some time of incubation probably due to the formation of inhibitory compounds such as polysulphides or changes in the thermodynamic conditions. The biotechnological application of such a slow process for wastewater treatment would only be feasible by inoculating the bioreactors with large amounts of seafloor sediment (which is impractical) or by enriching the AMNE/SR consortium in bioreactors by feeding it alternative substrates prior to feeding it methane (Meulepas et al., 2010b).

5.4.3 ACETATE AS A KEY INTERMEDIATE IN THE AOM-SR

Similarly to the results presented in the present work, acetate addition did not inhibit AOM as would be expected from an intermediate for AOM-SR in the work of Orcutt et al., (2008). In fact, the microbial activity was stimulated by this addition. It is unlikely that another SRB group was present in the inoculum to account for the acetate oxidation as this would mean a very rapid activation of the latter.

In addition, the fact that acetate production and consequent oxidation was observed in the groups where no acetate was added gives stronger evidence for acetate playing a key role in the process. However, this work does not exclude the possibility of more than one intermediates being involved in the process as has been hypothesized in other publications (Valentine et al., 2000; Stams and Plugge, 2009).

Our observations show that sulphide does not account for all of the sulphate reduced in the batch tests. Theoretically sulphide should have been 10x higher. One of the other end products detected was thiosulphate. However, it is not clear via what pathway thiosulphate is formed. The amount of thiosulphate produced in the groups where acetate was added is lower than in the other groups where sulphate reduction is substantial. Thus, the presence of acetate might contribute to a more complete reduction of sulphate. Although not measured, it is also possible that disulphides or elemental sulphur were formed as was proposed in other publications (Milucka et al., 2013).

In this work, we propose a model for AOM-SR using acetate as an intermediate (Figure 5.8). In the latter, ANME utilizes methane and carbonate to produce acetate which is then oxidized by the consortial partner SRB to reduce sulphate. From the latter, sulphide, carbonate and thiosulphate (and maybe other S compounds) are produced. Note that thiosulphate may be formed independently from this specific pathway). It is also hypothesized that ANME is capable of methanogenesis, utilizing acetate to produce methane and carbonate.

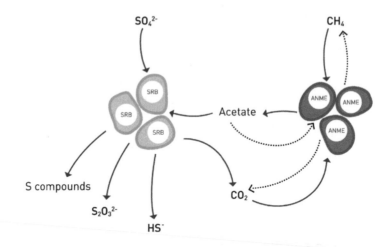

FIGURE 5.8 Proposed model for Anaerobic Oxidation Coupled to Sulphate Reduction.

5.5 CONCLUSION

· Sulphate reduction and anaerobic oxidation of methane show different growth rates probably due to other microbial processes taking place such as methanogenesis.

· Acetate appears to play a key role in the syntrophic consortia.

· Sulphate reduction does not lead only to sulphide production but also to the production of thiosulphate and probably other S compounds.

· AOM-SR consortia are able to have diverse pathways, being able to use sulphate, sulphur, thiosulphate and acetate.

· Microbial analyses currently being performed can assess the effects of these co-substrates on community shifts.

Supporting Information

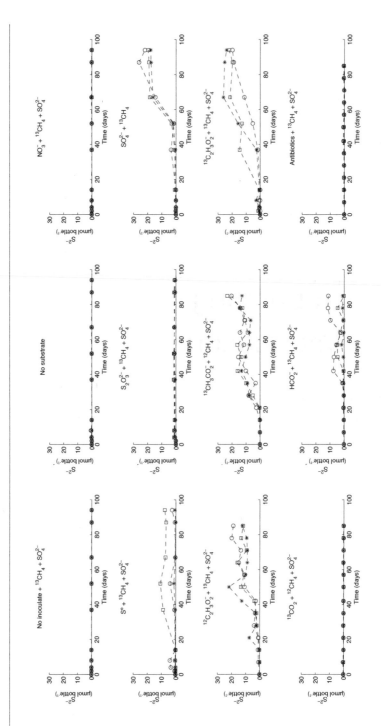

FIGURE 5S.1 Sulphide production in time in batch incubations in the absence (control) or presence of co-substrate.

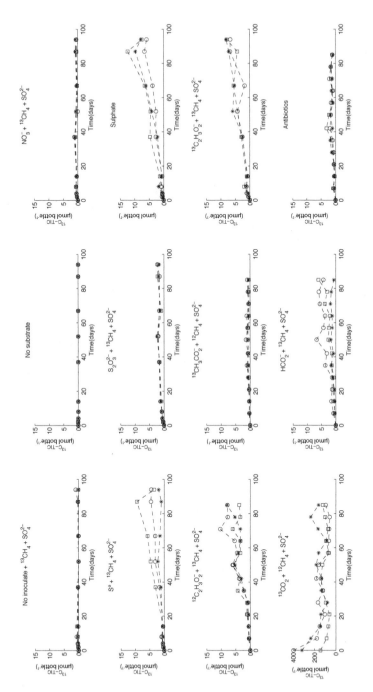

FIGURE 5S.2 ^{13}C-TIC production in time in batch incubations in the absence (control) or presence of co-substrate.

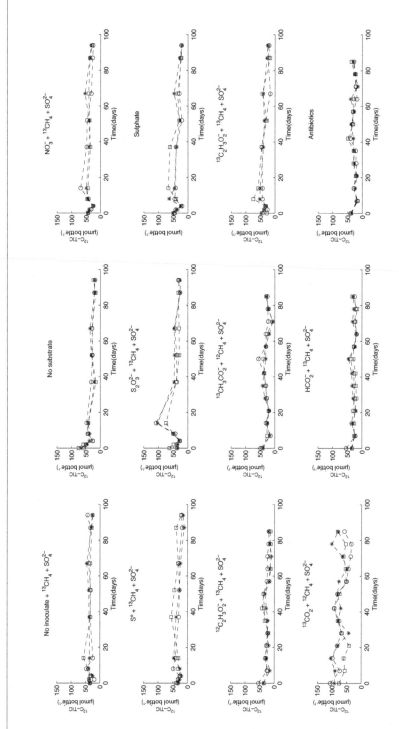

FIGURE 5S.3 ^{12}C-TIC production in time in batch incubations in the absence (control) or presence of co-substrate.

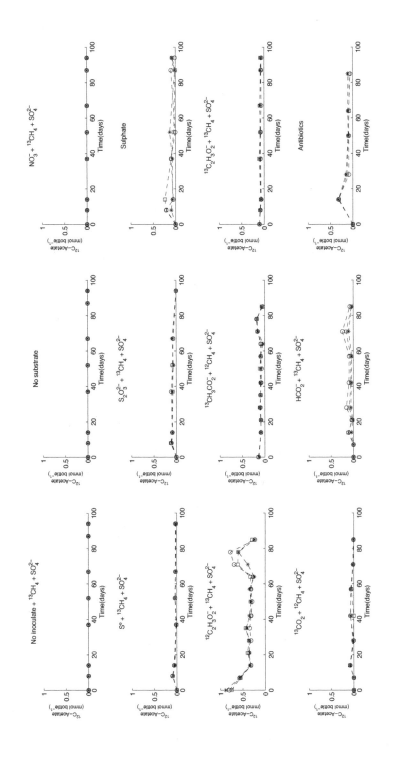

FIGURE 5S.4 ^{12}C-Acetate production in time in batch incubations in the absence (control) or presence of co-substrate.

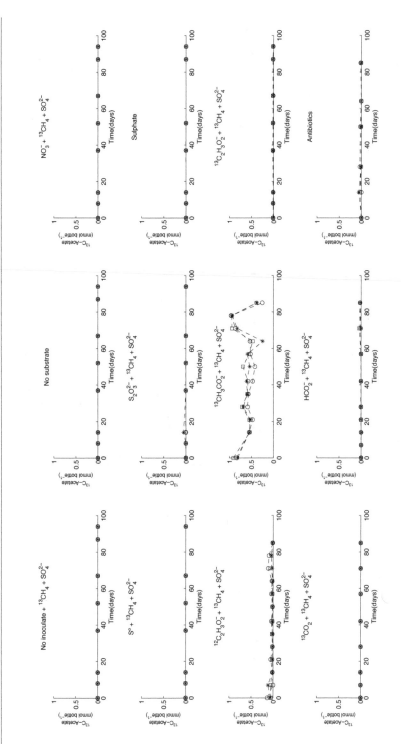

FIGURE 5S.5 ^{13}C-Acetate production/consumption in time in batch incubations in the absence (control) or presence of co-substrate.

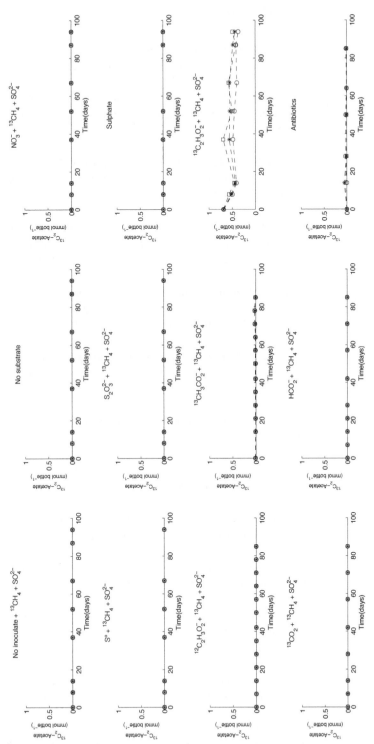

FIGURE 5S.6 $^{13}C_2$-Acetate production/consumption in time in batch incubations in the absence (control) or presence of co-substrate.

REFERENCES

Alperin MJ, Hoehler TM (2010) Anaerobic methane oxidation by archaea/sulfate-reducing bacteria aggregates: 1. Thermodynamic and physical constraints. Am J Sci 309: 869–957.

Beal EJ, Claire MW, House CH (2011) High rates of anaerobic methanotrophy at low sulfate concentrations with implications for past and present methane levels. Geobiology 9: 131–139.

Boetius A, Ravenschlag K, Schubert CJ, Rickert D, Widdel F, Gleeske A, Amann R, Jorgensen BB, Witte U, Pfannkuche O (2000) A marine microbial consortium apparently mediating AOM. Nature 407: 623–626.

Bowles MW, Samarkin V a., Bowles KM, Joye SB (2011) Weak coupling between sulfate reduction and the anaerobic oxidation of methane in methane-rich seafloor sediments during *ex situ* incubation. Geochim Cosmochim Acta 75: 500–519.

Duan Z, Mao S (2006) A thermodynamic model for calculating methane solubility, density and gas phase composition of methane-bearing aqueous fluids from 273 to 523K and from 1 to 2000 bar. Geochim Cosmochim Acta 70: 3369–3386.

Duan Z, Sun R (2003) An improved model calculating CO_2 solubility in pure water and aqueous NaCl solutions from 273 to 533 K and from 0 to 2000 bar. Chemical Geology 193: 257–271.

Duan Z, Sun R, Liu R, Zhu C (2007) Accurate thermodynamic model for the calculation of H_2S solubility in pure water and brines. Energy Fuels 91:2056–2065.

Hinrichs K, Hayes JM, Sylva SP (1999) Methane-consuming archaebacteria in marine sediments. Nature 398: 802–805.

Jørgensen BB, Kasten S (2006) Sulfur Cycling and Methane Oxidation. Marine Geochemistry, (Schulz HD & Zabel M, eds), pp. 271–309. Springer, Berlin.

Knittel K, Boetius A (2009) Anaerobic Oxidation of Methane: Progress with an unknown process. Annu Rev Microbiol 63: 311–334.

Meulepas, Jagersma CG, Zhang Y, Petrillo M, Cai H, Buisman CJN, Stams AJM, Lens PNL (2010a) Trace methane oxidation and the methane dependency of sulfate reduction in anaerobic granular sludge. FEMS Microbiol Ecol 72: 261–271.

Meulepas RJW, Stams AJM, Lens PNL (2010b) Biotechnological aspects of sulfate reduction with methane as electron donor. Rev Environ Sci Bio/Technology 9: 59–78.

Meulepas RJW, Jagersma CG, Khadem AF, Stams AJM, Lens PNL (2010c) Effect of methanogenic substrates on anaerobic oxidation of methane and sulfate reduction by an anaerobic methanotrophic enrichment. Appl Microbiol Biotechnol 87: 1499–1506.

Milucka J, Ferdelman TG, Polerecky L, Franzke D, Wegener G, Schmid M, Lieberwirth I, Wagner M, Widdel F, Kuypers MMM (2012) Zero-valent sulphur is a key intermediate in marine methane oxidation. Nature 491: 541–546.

Nauhaus K, Boetius A, Krüger M, Widdel F (2002) *In vitro* demonstration of anaerobic oxidation of methane coupled to sulphate reduction in sediment from a marine gas hydrate area. Environ Microbiol 4: 296–305.

Nauhaus K, Treude T, Boetius A, Krüger M (2005) Environmental regulation of the anaerobic oxidation of methane: A comparison of ANME-I and ANME-II communities. Environ Microbiol 7: 98–106.

Nauhaus K, Albrecht M, Elvert M, Boetius A, Widdel F (2007) *In vitro* cell growth of marine archaeal-bacterial consortia during anaerobic oxidation of methane with sulfate. Environ Microbiol 9: 187–196.

Orcutt B, Meile C (2008) Constraints on mechanisms and rates of anaerobic oxidation of methane by microbial consortia: process-based modeling of ANME-2 archaea and sulfate reducing bacteria interactions. Biogeosciences 5:1587–1599.

Orcutt B, Samarkin V, Boetius A, Joye S (2008) On the relationship between methane production and oxidation by anaerobic methanotrophic communities from cold seeps of the Gulf of Mexico. Environ Microbiol 10: 1108–1117.

Orphan VJ, Hinrichs K-U, Ussler III W, Paull CK, Taylor LT, Sylva SP, Hayes JM, Delong EF (2001) Comparative analysis of methane-oxidizing archaea and sulfate-reducing bacteria in anoxic marine sediments. Appl Environ Microbiol 67: 1922–1934.

Reeburgh WS (2007) Oceanic Methane Biogeochemistry. Chem Rev 107: 486–513.

Sørensen KB, Finster K, Ramsing NB (2001) Thermodynamic and kinetic requirements in anaerobic methane oxidizing consortia exclude hydrogen, acetate, and methanol as possible electron shuttles. Microb Ecol 42: 1–10.

Stams AJM, Plugge CM (2009) Electron transfer in syntrophic communities of anaerobic bacteria and archaea. Nat Rev Microbiol 7: 568–577.

Tang Y, Ontiveros-Valencia A, Feng L, Zhou C, Krajmalnik-Brown R, Rittmann BE (2013) A biofilm model to understand the onset of sulfate reduction in denitrifying membrane biofilm reactors. Biotechnol Bioeng 110: 763–772.

Valentine DL (2002) Biogeochemistry and microbial ecology of methane oxidation in anoxic environments: A review. Int J Gen Mol Microbiol 81: 271–282..

Valentine DL, Reeburgh WS, Hall R (2000) Minireview New perspectives on anaerobic methane oxidation. Environ Microbiol 2:477-484.

Wang FP, Zhang Y, Chen Y, He Y, Qi J, Hinrichs K-U, Zhang X-X, Xiao X, Boon N (2014) Methanotrophic archaea possessing diverging methane-oxidizing and electron-transporting pathways. ISME J 8: 1069–1078.

Widdel F, Bak F (1998) Gram-negative mesophilic sulphate-reducing bacteria. The Prokaryotes, (Balows A, Truper HG, Dworkin M, Harder W, Schleifer K-H, eds), pp. 3352–3378. Springer New York.

Zhang Y, Henriet JP, Bursens J, Boon N (2010) Stimulation of *in vitro* anaerobic oxidation of methane rate in a continuous high-pressure bioreactor. Bioresour Technol 101: 3132–3138.

6

General Discussion and Recommendations

6.1 INTRODUCTION

Several industrial activities are responsible for the release of sulphate and other sulphur compounds into freshwater streams. In addition, the levels of sulphate can be further increased due to the intrusion of seawater. Sulphate does not present an elevated environmental risk when compared to other pollutants. However, in addition to increasing the salinity in freshwater bodies, it can pose a great threat when, in the absence of oxygen and nitrate, it is reduced to hydrogen sulphide by sulphate reducing bacteria (SRB). Hydrogen sulphide causes an unpleasant smell, corrosion problems, may lead to mobilization of heavy metals and is highly toxic causing death at gaseous concentrations above 800-1000 mg.L^{-1} (Speece, 1996).

Diverse treatment methods can be applied to sulphate containing wastewaters. Such techniques can be membrane filtration and chemical methods (e.g., adsorption and filtration) which are expensive, and require a post-treatment of the brine. Thus the most cost-effective removal of sulphate from wastewater is the biological sulphate removal by anaerobic bacteria (Lens et al., 1998). In addition, the end product of biological reduction of sulphate is hydrogen sulphide which can be used for the removal of heavy metals from wastewater. Several bioreactor configurations have been designed for biological sulphate reduction, which aim at increasing biomass retention to compensate for the low growth rates of the anaerobic microorganisms. These include batch reactors, sequencing batch reactors, continuously stirred tank reactors, anaerobic contact processes, anaerobic baffled reactors, anaerobic filters, fluidized bed reactors (up-flow and down-flow), gas lift reactors, up-flow anaerobic sludge blanket reactors, anaerobic hybrid reactors and membrane bioreactors (Kaksonen and Puhakka, 2007).

SRB require the presence of an electron donor and can be divided in two main subgroups: heterotrophic and autotrophic SRB. Heterotrophic SRB use organic matter and autotrophic SRB utilize carbon dioxide and hydrogen as electron donors (Liamleam and Annachhatre, 2007). Industrial wastewater is usually deficient in organic matter and thus need to be supplemented with appropriate electron donors. The choice in electron donor must rely of the following aspects: price, availability, residual colour or pollution, suitability for a specific waste or process water (volume, composition and salinity) and legislation regarding safety and environment (Bijmans, 2008).

The process can be further optimized to reduce costs. As the electron donor is one of the major costs for SR, this research focused on how to optimize its input. This work investigated two different approaches to optimize biological sulphate reduction: to develop a process control strategy to optimize the input of an electron donor and to increase the feasibility of a cheap carbon source, i.e., methane.

Most research on anaerobic oxidation of methane coupled to sulphate reduction (AOM-SR) has focused on *in situ* conditions and only few have studied the possibility to apply this process to wastewater treatment technologies. To use methane as an electron donor is highly attractive given the advantages depicted in Figure 6.2. The greatest bottleneck for the biotechnological application of AOM-SR is the extremely low growth rate of the responsible microorganisms, with doubling time up to 7 months (Nauhaus et al., 2007).

Thus, for technological applications the microbial activity must be maximized so that the enrichment can perform high rate sulphate removal. For this, research should first focus on understanding the pathways taking place, e.g., gaining knowledge on which are the key intermediates in the syntrophic relationship between anaerobic methanotrophs and SRB. On one hand, high pressure incubation procedures are not appealing for full-scale applications given their high energy requirements but on the other hand they seem to be the right choice to gain knowledge on the process since the biomass responsible for AOM-SR originates from deep sea and thus, from high pressure conditions.

Bioprocess monitoring and control is crucial to help maintain the ideal conditions for microorganisms while optimizing the process goals. Automatic control of the sulphide production can avoid excess dosage of electron donor which can pose increased costs and excess carbon in the effluent. This is especially interesting for processes involving the removal of heavy metals by precipitation with sulphide (Villa-Gomez et al., 2014). Furthermore, bioprocess control in sulphate reducing bioreactors can be used to regulate the competition between different organisms which can play a great role in electron donor consumption. In such systems the process dynamics will diversify with time and thus, it is advisable to resource to adaptive control where the control parameters are adjusted to the process dynamics (Cassidy et al., 2015). Validated mathematical models can be of great use for such control strategies as they can simulate and predict the variations in the system. Even though, the models developed so far for biological sulphate reducing processes (Tables 2.1-2.4; Chapter 2) were not meant for adaptive control strategy for optimizing the input of electron donor, they can and should be used as a starting point for developing such strategies.

This chapter summarizes and discusses the implications of the main findings of this thesis for the optimization of electron donor for sulphate reducing systems. In addition, it presents recommendations for future research.

6.2 OBJECTIVE

The overall objective of this research was to evaluate different ways to optimize biological sulphate reduction. Specifically, two approaches were considered:

1. To study the different stages in developing a process control strategy to optimize the input of a commonly used electron donor, i.e., lactate.

2. To get further insight on how to increase the feasibility of using a cheap carbon source such as methane by gaining knowledge on the pathways in AOM-SR;

The first approach consisted in studying different tuning strategies, evaluating the use of a pS electrode and developing a model of the sulphate reduction process with an accumulation factor in feast/famine behaviour. The second approach was to study the effect of different co-substrates on the AOM and SR rates in high pressure incubations. These objectives and main research findings are depicted in Figure 6.1 and Figure 6.2.

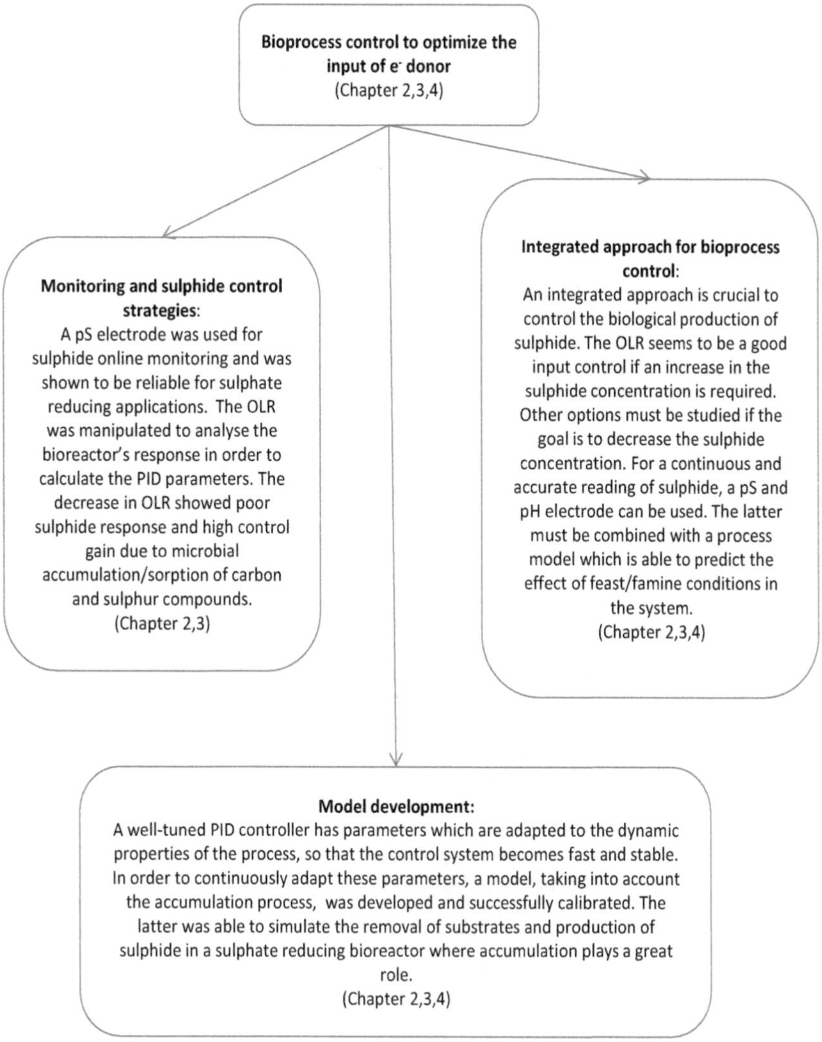

FIGURE 6.1 Summary of the main results of this thesis related to the development of a bioprocess control for biological sulphate reduction. PID: proportional, integrative, derivative; OLR: organic loading rate.

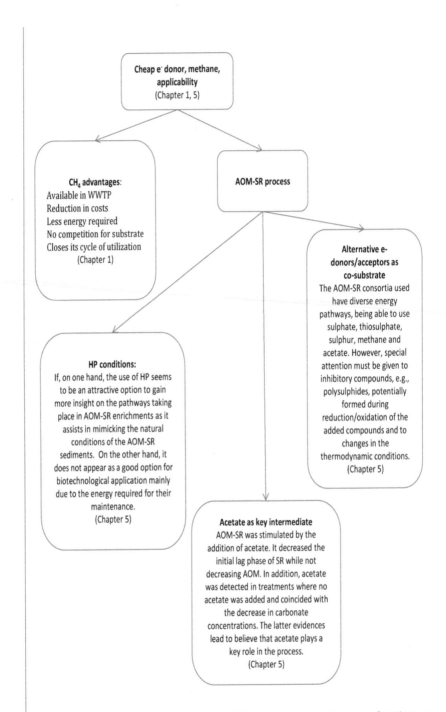

FIGURE 6.2 Summary of the main results of this thesis related to the usage of methane as a carbon source for biological sulphate reduction. HP: High pressure; AOM-SR: Anaerobic oxidation of methane coupled to sulphate reduction; WWTP: Wastewater treatment plant.

6.3 BIOPROCESS CONTROL FOR BIOLOGICAL SULPHATE REDUCING PROCESSES

For the full scale application of sulphur cycle based biotechnologies, it is of crucial importance to design and implement efficient control strategies to optimize the microorganisms' growth and competition, to control inhibitory factors and/or to optimize the production of products for secondary processes, e.g., heavy metal precipitation with sulphide (Cassidy et al., 2015). The design of a successful control strategy is not an easy task as it involves the understanding of the process dynamics. This thesis presented the first crucial steps to develop such a control strategy (Chapter 3 and 4).

After defining the goal(s) for the process control, the first step should be the selection of an appropriate monitoring device of the variable(s) to control. Sensors applicable to sulphate reducing bioreactors have been extensively reviewed (Cassidy et al., 2015; Tables 2.9 - 2.10). In this research a solid state Ag_2S ion selective pS electrode was validated for the online monitoring of sulphide (Figure 3.2-3.3; Chapter 3). The use of a pS electrode in bioreactors presents a great advantage over offline methods as it avoids losses due to volatilization or oxidation (Hu et al., 2010) and maximizes analysis speed. It must be noted that due to speciation of sulphide, pH measurements must be taken simultaneously in order to calculate total dissolved sulphide more accurately utilizing equation 3.1 from Chapter 3 of the current thesis.

The second step is to define the control input, i.e., which variable to manipulate in order to achieve the defined goal. This study showed that varying the organic loading rate (OLR) created sulphide responses suitable for calculating the process control proportional-integral-derivative (PID) parameters (Table 3.2; Chapter 3). The OLR was manipulated by changing the COD_{in} the influent and by changing the hydraulic retention time (HRT) and showed adequate response when the aim was to increase the production of sulphide with the change in COD being the most effective strategy (Figure 3.5; Chapter 3). Despite the adequate sulphide response, several days (response time) were needed to achieve pS steady state values in both tuning strategies that can destabilize the bioreactor in case of excessive control actions. This can lead to a COD overload and hence, substrate inhibition (Qatibi et al., 1990) or sulphide toxicity (Reis et al., 1992) when applying a change in COD_{in} concentration or biomass washout (Kaksonen et al., 2004) when applying a change in the HRT.

However, decreasing the OLR did not cause the pS values to decrease as was initially expected (Figure 3.5; Chapter 3). Thus, it was hypothesized that the feast/famine conditions stimulated the accumulation of storage carbon and sulphur compounds. SRB and other microorganisms are capable of accumulating such compounds (Cypionka, 1989; Hai et al., 2004) and to our knowledge, this is the first time that such storage is studied for continuous systems treating sulphate containing wastewaters. Evidence for this accumulation is shown in Chapter 4 through several shock loads. When the addition of COD and sulphate was stopped, sulphide was still being produced after 15 days of operation (Figure 4.3a; Chapter 4). It is hypothesized that accumulation of polyhydroxybutyrate (PHB) (Figure 4.2; Chapter 4) and the accumulation/sorption of sulphate were the sources for the continuous production of biological sulphide. The production of sulphide ceased when the concentration of PHB was decreased to zero (Figure 4.3b; Chapter 4). The system responded quickly to a one day COD load producing sulphide (Figure 4.3c; Chapter 4). Thus, the activity of microorganisms did not seem to be greatly affected by the shock loads induced in this study. In fact, the accumulation of storage compounds allows the microorganisms to maintain a balanced metabolism under limited substrate conditions and in different feed shocks (van Loosdrecht et al., 1997).

Given the great dynamics of this process discussed in the previous paragraphs, the third step of the control strategy should be the development of a model capable of simulating and predicting the various pathways that are simulated by the different changes in the operational conditions.

Chapter 4 presents for the first time a model for sulphate reduction which accounts for both polyhydroxybutyrate (PHB) and sulphate accumulation. The model was calibrated and showed a very good fit between simulated and experimental data for the COD, sulphate and sulphide variables in the effluent of an inversed fluidized bed reactor (Figure 4.6; Chapter 4). This model can assist in understanding how the different shock loads will have an impact on the SRB activity so that excessive control actions which can destabilize the reactor can be avoided.

The design of a successful control strategy is not an easy task as it involves the understanding of the process dynamics (Cassidy et al., 2015). The work presented in this thesis has helped to define three crucial steps in the control strategy development for sulphide control in sulphate reducing bioreactors. The main steps for control strategy development are shown in Figure 6.3.

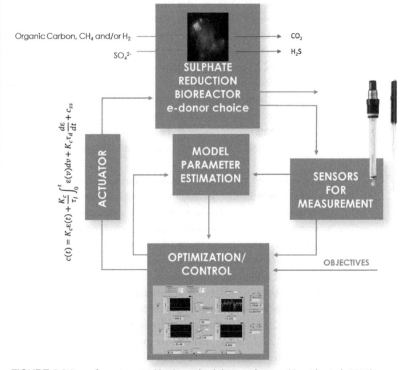

FIGURE 6.3 Steps for automated biological sulphate reduction (Cassidy et al., 2015).

6.4 METHANE AS ELECTRON DONOR FOR SULPHATE REDUCTION – CHALLENGES FOR BIOTECHNOLOGICAL APPLICATION

For biological sulphate reduction processes, the choice of electron donor presents a big effect on the overall cost of the process. As such, methane appears to be a good alternative to the currently utilized electron donors (Table 1.1; Chapter 1). However, sulphate reduction with methane should present high sulphate removal rates at ambient pressure which have not been achieved so far. This research has tried to obtain more knowledge on the pathways taking place in the consortia responsible for AOM-SR. For the latter, efforts were focused on adding several labelled and non-labelled substrates to high pressure *in vitro* batch incubations. The predicted 1:1 ratio commonly accepted for AOM-SR was not observed in this work as both the sulphide production was 3 times higher (Figure 5.2; Chapter 5) and sulphate reduction was 50 times higher (Figure 5.3; Chapter 5) than what would be expected theoretically for the total inorganic carbon produced (Figure 5.4; Chapter 5). These results are in agreement with other studies which show that the 1:1 ratio is rarely observed (Beal et al., 2011; Bowles et al., 2011).

These results support the hypothesis for other microbial processes occurring simultaneously, i.e., methanogenesis. Several studies have shown that it can be performed by ANME accounting for around 10% of the total AOM rate (Orcutt et al., 2008; Meulepas et al., 2010a; Bowles et al., 2011). In addition, all genes required for methanogenesis were present and actively expressed in the ANME 2a used for the experiments performed in this study (Wang et al., 2014).

The used enrichment was shown to have diverse energy pathways, being able to use sulphate, thiosulphate, sulphur, formate and acetate. Although the enrichment was capable of utilizing thiosulphate, sulphur and formate, the AOM and SR rates showed lower rates probably due to the production of inhibitory compounds (Figure 5.2 and Figure 5.4; Chapter 5). Acetate, however, was not only used by the enrichment, but also enhanced the SR process by decreasing the initial lag phase of the sulphide production (Figure 5.2; Chapter 5). It is important to note that it is unlikely that any other SRB group would be responsible for the acetate oxidation since it would mean its very rapid activation as no acetate was added throughout the several years of previous incubation of the sediment. Acetate has been shown to be a thermodynamically favourable intermediate of AOM-SR process when high methane concentrations are present (Valentine, 2002; Strous and Jetten, 2004) as is the case in the experiments performed in this research. In addition, the fact that acetate production and consequent oxidation was observed in the groups where no acetate was added gives stronger evidence for acetate playing a key role in the process (Figure 5.5; Chapter 5). A process model is proposed based on these findings and depicted in Figure 6.4. However, this work does not exclude the possibility of more than one intermediate being involved in the process as has been hypothesized in other publications (Valentine et al., 2000; Stams and Plugge, 2009). Thus, the possibility to use acetate to enhance the rates for AOM and SR must be studied further in order to be applied for wastewater treatment.

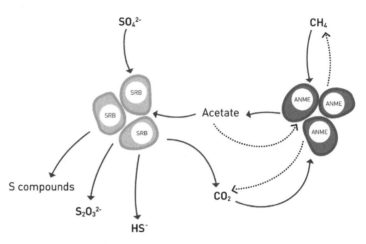

FIGURE 6.4 Proposed model for AOM-SR with acetate as key intermediate.

6.5 CONCLUSIONS AND RECOMMENDATIONS FOR FUTURE RESEARCH

The cost related to the dosage of electron donor can be optimized by the two approaches studied in this research. However, further research is needed to allow full-scale applications.

6.5.1. ADAPTIVE BIOPROCESS CONTROLLER FOR THE CONTROL OF SULPHIDE PRODUCTION

In practice, bioprocess control for anaerobic processes treating wastewater is not an easy task as it involves full understanding of the process dynamics. There are three main aspects to consider when developing an adaptive process control: Control strategy (control input and output, controller type), Monitoring (choice of sensors) and Modelling (understand and predict process dynamics).

For this research the OLR was chosen as a control input. Although a change in the influent COD gave satisfying results for when the goal was to increase the sulphide, it did not do the same for the decrease in sulphide concentration (Figure 3.4; Chapter 3) given microbial accumulation processes (Figure 4.2-4.3; Chapter 4). Thus, other control inputs must be considered when a decrease in sulphide concentration in IFB bioreactor is desired. Such control inputs might be to dilute the effluent (Metcalf and Eddy, 2002) or to sparge the reactor with oxygen free nitrogen gas. The latter seems to be the best as it does not lead to substrate losses (in case of liquid electron donors) as diluting the system would. The several induced shock loads, necessary for control purposes, might have a great effect on the microbial competition and thus, close attention must be given to this. In sulphate reducing systems, competition between complete oxidizing SRB, incomplete oxidizing SRB, methanogens, fermenters and acetogens play a great role (Vavilin et al., 1994; Kalyuzhnyi and Fedorovich, 1998; Lens et al., 1998; Omil et al., 1998; Frunzo et al., 2012). To control the outcome of this competition, one might consider including other online sensors to measure, for example, methane and acetate.

One of the major drawbacks encountered for the control of sulphide production is the long response time of the system to step changes (Table 3.2; Chapter 3). Another adaptation of the PID parameters can be obtained from the information of the dynamics of the bioprocess such as reaction pathways and kinetics as well as mass balances, which are always required in biological systems due to their non-linearity and non-stationary characteristics (Steyer et al., 2000). This information can help to predict the response time of the system to the applied change thus preventing an excessive control action. In addition, a maximum pH and COD input must be defined to avoid inhibition.

The model developed and calibrated in Chapter 4 of this thesis is a great starting point for a model to be used in an adaptive control strategy as it is able to simulate the removal and production processes in the reactor taking into account the accumulation processes that caused such long time responses previously mentioned. The model should be further developed to be able to automatically calculate the PID parameters and thus adapt the parameter values in the actuator (Figure 6.3). Furthermore, if the ultimate goal is heavy metal recovery by metal sulphide precipitation, the kinetics for such processes should also be included in the model.

6.5.2 ALTERNATIVE ELECTRON DONORS FOR BIOLOGICAL SULPHATE REDUCTION

This research studied the feasibility of using methane as a carbon source for sulphate reduction (Chapter 5). The greatest bottleneck of this process for biotechnological application is the slow growth of the responsible microorganisms. Growth on alternative substrates might be a way to improve this growth (Meulepas et al., 2010b). As such, this research aimed at analysing which could be the possible substrates. The results show a diversity of possible energy pathways for the microorganisms performing AOM-SR. Acetate appeared as the most suitable and results even indicated the possibility of acetate being an intermediate of the process (Figure 6.4). Consequently, it is advised to further study the use of acetate to understand if the microorganisms producing/consuming acetate are the same microorganisms performing AOM-SR and if so enrich AOM-SR performing communities. In addition, the possibility of using more than one additional substrate should also be considered.

The isolation of the responsible microorganisms for AOM-SR has not been achieved. Even though it is a difficult task, it would not only allow to gain detailed knowledge on their physiology, behaviour and interactions with other organisms (Muyzer and Stams, 2008), but also to help understand better how to stimulate the process for feasible biotechnological application.

The experiments for AOM-SR in this thesis were performed in batch tests. The same inoculum has been incubated and enriched in a high pressure continuous system (data not shown) and different operational conditions caused great impact in the prevailing microbial pathways. Thus, it is suggested to study the response of such consortia to these changes and also to feed the reactor with co-substrates such as acetate to obtain faster growth rates.

High pressure conditions are advisable to study these communities as they mimic the natural conditions of the latter. However, it is not feasible for biotechnological applications given their high energy and safety requirements. Hence, the development of ambient pressure bioreactors which account for high biomass retention are of extreme importance.

Besides methane, slow release electron donors could also be considered, e.g, PHB. PHB is produced by the bacterial community in the bioreactor as shown in Figure 4.3 in Chapter 4. Thus, research should focus on how to optimize this production by, e.g., providing feast/famine conditions, and so decrease the use of external electron donors such as lactate. The model developed in Chapter 4 can be used as a support for such studies.

REFERENCES

Beal EJ, Claire MW, House CH (2011) High rates of anaerobic methanotrophy at low sulfate concentrations with implications for past and present methane levels. Geobiology 9: 131–139.

Bijmans MFM (2008) Sulfate reduction under acidic conditions for selective metal recovery. Wageningen University, Wageningen, The Netherlands.

Bowles MW, Samarkin VA, Bowles KM, Joye SB (2011) Weak coupling between sulfate reduction and the anaerobic oxidation of methane in methane-rich seafloor sediments during ex situ incubation. Geochim Cosmochim Acta 75: 500–519.

Cassidy J, Lubberding HJ, Esposito G, Keesman KJ, Lens PNL. (2015) Automated biological sulphate reduction: a review on mathematical models, monitoring and bioprocess control. FEMS Microbiol Rev. 39(6):823-53

Cypionka H (1989) Characterization of sulfate transport in *Desulfovibrio desulfuricans*. Arch Microbiol 152:237–243.

Frunzo L, Esposito G, Pirozzi F, Lens P (2012) Dynamic mathematical modeling of sulfate reducing gas-lift reactors. Process Biochem 47: 2172–2181.

Hai T, Lange D, Rabus R, Steinbüchel A (2004) Polyhydroxyalkanoate (PHA) accumulation in sulfate-reducing bacteria and identification of a class III PHA synthase (PhaEC) in *Desulfococcus multivorans*. Appl Environ Microbiol 70: 4440–4448.

Hu P, Jacobsen L, Horton J, Lewis R (2010) Sulfide assessment in bioreactors with gas replacement. Biochem Eng J 49: 429–434.

Kaksonen AH, Puhakka J A(2007) Sulfate reduction based bioprocesses for the treatment of acid mine drainage and the recovery of metals. Eng Life Sci 7: 541–564.

Kaksonen AH, Franzmann PD, Puhakka JA (2004) Effects of hydraulic retention time and sulfide toxicity on ethanol and acetate oxidation in sulfate-reducing metal-precipitating fluidized-bed reactor. Biotechnol Bioeng 86: 332–343.

Kalyuzhnyi S V, Fedorovich VV (1998) Mathematical modelling of competition betweent sulphate reductions and methanogenesis in anaerobic reactors. Bioresour Technol 65: 227–242.

Lens PNL, Visser A, Janssen AJH, Hulshof Pol LW, Lettinga G (1998) Biotechnological treatment of sulfate-rich wastewaters. Crit Rev Env Sci Technol 28: 41–88.

Liamleam W, Annachhatre AP (2007) Electron donors for biological sulfate reduction. Biotechnol Adv 25: 452–463.

Metcalf, Eddy (2002) Wastewater Enginering: Treatment and Reuse. McGraw-Hill, ed.

Meulepas RJ, Jagersma CG, Zhang Y, Petrillo M, Cai H, Buisman CJ, Stams AJ, Lens PNL (2010a) Trace methane oxidation and the methane dependency of sulfate reduction in anaerobic granular sludge. FEMS Microbiol Ecol 72: 261–271.

Meulepas RJW, Stams AJM, Lens PNL (2010b) Biotechnological aspects of sulfate reduction with methane as electron donor. Rev Environ Sci Bio/Technology 9: 59–78.

Muyzer G, Stams AJ (2008) The ecology and biotechnology of sulphate-reducing bacteria. Nat Rev Microbiol 6: 441–454.

Nauhaus K, Albrecht M, Elvert M, Boetius A, Widdel F (2007) In vitro cell growth of marine archaeal-bacterial consortia during anaerobic oxidation of methane with sulfate. Env Microbiol 9: 187–196.

Omil F, Lens P, Visser A, Hulshof Pol LW, Lettinga G (1998) Long-term competition between sulfate reducing and methanogenic bacteria in UASB reactors treating volatile fatty acids. Biotechnol Bioeng 57: 676-685.

Orcutt B, Samarkin V, Boetius A, Joye S (2008) On the relationship between methane production and oxidation by anaerobic methanotrophic communities from cold seeps of the Gulf of Mexico. Environ Microbiol 10: 1108–1117.

Qatibi AI, Bories A, Garcia JL (1990) Effects of sulfate on lactate and C2- , C3- volatile fatty acid anaerobic degradation by a mixed microbial culture. Antonie Van Leeuwenhoek 58: 241–248.

Reis MAM, Almeida JS, Lemos PC, Carrondo MJT (1992) Effect of hydrogen sulfide on growth of sulfate reducing bacteria. Biotechnol Bioeng 40: 593–600.

Speece RE (1996) Anaerobic biotechnology for industrial wastewaters. Archae Press, Tennessee.

Stams AJM, Plugge CM (2009) Electron transfer in syntrophic communities of anaerobic bacteria and archaea. Nat Rev Microbiol 7: 568–577.

Steyer J, Buffière P, Rolland D, Moletta R (2000) Advanced control of anaerobic digestion processes through disturbances monitoring. Water Res 33: 2059–2068.

Strous M, Jetten MSM (2004) Anaerobic oxidation of methane and ammonium. Annu Rev Microbiol 58: 99–117.

Valentine DL (2002) Biogeochemistry and microbial ecology of methane oxidation in anoxic environments: A review. Antonie van Leeuwenhoek, Int J Gen Mol Microbiol 81: 271–282.

Valentine DL, Reeburgh WS, Hall R (2000) Minireview New perspectives on anaerobic methane oxidation. Environ Microbiol 2: 477-484.

van Loosdrecht M, Pot M, Heijnen J (1997) Importance of bacterial storage polymers in bioprocesses. Water Sci Technol 35: 41–47.

Vavilin VA, Vasiliev VB, Rytov S V, Ponomarev A V (1994) Self-oscilating coexistence of methanogens and sufate-reducers under under hydrogen sulfide inhibition and the pH-regulating effect. Biores Technol 49: 105–119.

Villa-Gomez DK, Cassidy J, Keesman KJ, Sampaio R, Lens PNL (2014) Sulfide response analysis for sulfide control using a pS electrode in sulfate reducing bioreactors. Water Res 50: 48–58.

Wang F-P, Zhang Y, Chen Y, He Y, Qi J, Hinrichs K-U, Zhang X-X, Xiao X, Boon N (2014) Methanotrophic archaea possessing diverging methane-oxidizing and electron-transporting pathways. ISME J 8: 1069–1078.

T - #0666 - 101024 - C0 - 244/170/10 - PB - 9781138029507 - Gloss Lamination